Train to Win

By

Wes Doss

This book is a work of non-fiction. Names and places have been changed to protect the privacy of all individuals. The events and situations are true.

ISBN: 1-4107-0198-0 (E-book)
ISBN: 1-4107-0199-9 (Paperback)

This book is printed on acid free paper.

1stBooks – rev02/04/03

Table of Contents

<u>Dedication</u>

To my best friend, my life long love, my soul and my inspiration, my wife Hye Chong and to my two greatest heros whom I admire for their courage, bravery and their discipline, my daughters Angela and Victoria. I dedicate this book with all my love. Without their support and inspiration, I could not have done it, nor would I be where I am today without them.

Forward

Never before in our history has there been such an interest in the study of self-defense, both armed and unarmed...and with good reason. Likewise, never has there been such a vast amount of information available to the hungry student. Unfortunately, with this availability comes a large amount of useless information penned by unqualified people whose sole aim is to make a dollar. On a brighter note, there is also a relatively smaller amount of information produced by professionals that gets to the meat and marrow of the matter at hand.

In this book, Mr. Wes Doss gets to the heart of the matter of self-protection. Wes is not only a tried professional, having served both in military and police uniforms, but he is also a gifted and articulate educator. Unlike many of his contemporaries, Wes has actual operational experience from which to draw his written lessons. His book goes beyond the typical "this is your front sight" instructional tomes, and trains the mind in addition to the hand.

I often decline when asked to write editorials and to review books. When Wes asked me to review his book and provide a foreword, I was both honored to do so, and pleased to read a refreshing and informative look into the subject of prevailing in personal combat. This book is a "must have" for all serious students. What you read herein, may very well keep you alive.

Gabe Suarez

Suarez International, Inc.

Introduction

Law enforcement officers, military personnel and other armed professionals must be thoroughly trained; not only in the basic function of their job description, but also in the detailed, specialized and unique skills that allow for maximum flexibility and adaptability to a wide array of situations. The world has grown into a substantially more complex place than just even ten years prior. The various statistics compiled by State and Federal agencies tell us that our violent crime rate has gone down considerably and steadily for years, while this may be true, the complexity and finesse of those perpetrating these acts has gone up. This results in a lowered rate, but an increase in the effectiveness of the act. This combined with the various policy changes, restrictive regulations, bizarrely written SOP's and concepts of community based programs has made the armed professional's job more difficult and much more precarious.

The world holds many instructors and training organizations that cater to the general and specific needs of the armed professional. A number of these training authorities are dedicated quality sources of knowledge and experience, providing a great product in a format that delivers accurate information that the customer can understand and apply. However, there is a disturbing number of organizations, many considered prominent, who are not providing such a product. Driven solely by the almighty dollar these individuals are concocting programs and techniques that are flashy and expensive, but are not relevant or practical. In fact, a great number of these celebrated figures have never been in a high-risk situation, nor understand the complications and after effects of the encounter.

The concept of enhancing human performance is not a new one as the research dates back to the 19th and late 18th century. However, very little of the research has been directly related to enhancing the performance of force or violence application. The majority of the study has been done in the fields of adult learning and sports performance. While these topics can seem quite separate from the needs of the armed professional they, really are not.

In the following pages, I have highlighted my personal beliefs in the field of force application training and the process required to break students out of a survival mindset and instill in them a winning mindset. The principles and theories that I discuss are provided simply as a guideline for instructors and will not guarantee success. However, I firmly believe in their value and that if used correctly, can assist greatly in the quest for winning warriors. Many of the situations I mention, my personal ones and those of my fellow and fallen colleagues, came about through a trial by fire type of circumstance. Many, in pursuit of a bigger and better technique, have paid the ultimate price while others, like myself, were lucky enough to walk away with our tail between our legs and visit to the hospital.

Many of the techniques or tactics in this book are not targeted at a specific situation. I did that for the purpose of making the reader, and hopefully other instructors, think. The whole concept behind training really hubs around rigidly flexible creativity. I believe firmly in the concept of using what works and cheating at all times in a fight. I see nothing more counter productive than the practice of creating carbon copy clones and thinking that the world is a "one size fits all" domain. The result, I believe, is a comprehensive look at teaching and training principles that will allow the instructor or administrator to develop a better and stronger program.

I did my very best to keep *Train to Win* as user friendly as I could. Generally, each chapter has overlapping themes and often I refer back to previous chapters in order to show the inter-relation of concepts and principles. I have tried to emphasize the concept of creating confidence, developing reflexive actions and the role that various principles of adult learning play in the force application-training arena. Further, I have tried to show the relationship between adult learning and the various reactions our own bodies have when encountering a high stress situation and how training the right way can mitigate those effects.

Train to Win is my effort to give back to those individuals for whom I hold the highest regard and who deserve the very best that we professional instructors can give. The men and women of the various law enforcement agencies that protect us day in and day out and those who have followed the calling of our armed forces and work diligently to keep our homeland and our interests safe and secure. These individuals have made that selfless, voluntary step into the professional world of force application and control. Our current state of world affairs has these brave men and women spread thin all over the globe and placed in a position where they must maintain a constant state of readiness. This state of readiness means that they need, no they deserve, the absolute highest quality of training that they can get, because their very existence is a thin line that provides us our sense of safety.

Train to Win was my absolute honor to compile and I dedicate it to these individuals.

Thank you,

Wes Doss

Acknowledgements

I would like to thank a number of very special men who stand as true warriors and academic professionals in the field of training. Each of them has contributed to the completion of this book in one form or another. I have personal relationships with many of them, while others I have studied from afar and have followed throughout their career. Specifically by name I want to thank the following:

Gabe Suarez	Ken Vogel	Bruce Bombeck
Bob Jackson	Louis Awerbuck	Robbie Barkman
Chris Sheppard	Ken Good	Bob Weber
Kevin McClung	Joe Ferrera	Robert Taylor
Roy Maas	Denny Hansen	Rich Lucibella
Shawn Brooks	Bodie Runston	John Clark

For there special help and creativity with the various photographs and illustrations in this book I want give special thanks to Mark Burnell of SG Five and Perry Taylor of Creative Concepts photography and for there moral and product support I want to thank the staff at Wiley-X eyewear.

Additionally, I must thank all those who I have stood behind, beside, and in front of. Those who I have gone through doors with and into foreign lands, those who I have shared the experience of being on the edge. Too numerous to list, you all know who you are and you are forever in my heart.

<u>Chapter One</u>

"I have never let my schooling interfere with my education"

-Mark Twain

<u>The Real World</u>

Throughout history there have always been individuals who sought out professions that placed them in harms way. Be it soldier, protector, guardian, defender, or lawbreaker the nature of these professions has called for a better than average understanding of weapons, tactics and strategy. As time went on, these professions shifted and changed to meet the ever evolving needs, as well as the ever-discursive side of society.

Not all professions engaging in martial activities have been honorable. Depending on your philosophical beliefs, if you believe in yin and yang or antagonistic opposites of good and evil, each member of society will view this separation differently. Each individual, depending on environmental, economic and societal conditions, forms different conclusions about what is good and what is evil. In many areas the police and even schoolteachers are viewed with condescension and hostility, while the gangsters and drug dealers are viewed as heroes. To some this seems like a bizarre phenomenon, but in reality it is quite common. The drug dealer is a sign of security, prestige and relative prosperity and the police and teachers are simply seen as the enemy, threatening an accustomed lifestyle. Regardless of your personal views of who is the good guy and who is the bad guy, it is important to acknowledge the division.

As history progressed, countless advancements in technology and information have contributed to the betterment of all aspects of supporting and training those martial occupations.

Those of us involved in training have a superabundance of information and aides to assist in our quest for more productive training with greater realism and more predictable results. With the onslaught of the 21st century, we find ourselves at a juncture where it is possible to elevate an individuals shooting, fighting, decision-making and situational control skills to levels not possible in years past. However, soldiers and police officers continue to get injured and killed in situations here they shouldn't, leaving one question unanswered…why? It would be simple to lay blame for these tragedies on one of many situations and perhaps there is no one answer. However, we can continue to work on the problem refusing to accept this as inevitability.

Survival, a real definition

Since the mid-1960's there has been a progressive, often digressive, shift among U.S. law enforcement agencies, particularly at county and municipal levels. This shift has been towards joining a widespread agenda known commonly as "community oriented policing". This growing concept, originally designed to build public trust in law enforcement, create accessibility, and reduce fears, has morphed into an overused "buzz word" used to describe a plethora of issues that are literally of no concern to the officer on the street or to the citizens at large. Within this era of community policing, political correctness, and zero tolerance rules a great many agencies have grown overly concerned with public perception, agency image, and vain attempts to immunize themselves from civil liabilities. These issues have very little to do with the actual dynamics of a high risk situation, but many times they are the basis for developing "model" officers and "model" training programs.

Like Mother Earth herself, the world that we as soldiers, police, and armed professionals operate in can be divided into two distinctive hemispheres, the objective realm and the subjective realm. The subjective realm is that domain that is influenced by our perceptions through stories and visions supplied to us by the media, movies, and "reality based" television shows. In the subjective realm sensationalism and excessive force are a way of life, the villain is easily identified, and the hero seldom makes mistakes or gets hurt. The subjective realm is dominated by pure fantasy and produces an extremely unrealistic view of the real world. The objective realm is that province that is based on the real-life and often harsh realities of our world and our occupations. The objective world has shown us time and again that our hero's are many times victims and that mistakes are much more commonplace than is perfection. The objective realm is a difficult and complicated environment, abundant with rules, regulations, and restrictions that only the good guys must follow. Additionally, the various aspirations and idiosyncrasies of many administrative bodies has numerous professionals spending valuable time appealing to a constructed need to pacify a very small percentage of the population. An extremely small portion of our society that believes that those in protective roles should devote their entire time to resolving social issues instead of perfecting their trade. The popular media plays a significant role in making this squeaking mouse sound like a roaring lion, bringing these various special interests to the forefront of our lives. The objective realm receives further complication from a strange and often fatal state of affairs brought about through unusually written policy, inefficient leadership and insufficient or ineffective training. Flawed training is typically not obvious until after a tragic event occurs, at which time our society cries out for explanation and our various bureaucratic and litigious entities go forth on a modern version of the Salem witch-hunts, looking for someone to blame. Eventually, a change occurs in the training environment, but not soon enough and usually not to the point that is needed. I like to explicate this scenario on simple terms, so as you read imagine a large deflated balloon, a child's toy. In its deflated state it appears to be uniform in construction, with no visible flaws; as perfect as a balloon can be. Imagine now the balloon being inflated, reaching its maximum capacity, and an undetected weak spot becomes stressed beyond the balloons ability to expand and ultimately the balloon bursts. Like the flaw in the balloon, problems in training are not noticeable until it's too late. These flaws only rear their ugly heads at the worst times, manifesting themselves in the middle of a high stress situation.

Tragic, but true, that training flaws are not usually evident until it's too late.

(Courtesy of Perry Taylor)

Flawed training easily finds its way into nearly every facet of what is considered use of force training, largely because of the mentality of those who authorize or administrate training. Generally, the purpose and process of training are not well understood by the managers and administrators of most agencies. In fact, the majority of managers in protective fields don't have a formal or even functional background in either training or education. These individuals end up relying on their own misconception and opinions, or those of others, in order to make decisions about training. Too often, the senior administrators of an agency get no more involved in the training process, than overseeing the final approval of the budget, preferring to delegate all other responsibilities to someone else who may not be qualified for the task. Under these circumstances, agencies often end up adopting irrelevant and expensive training programs that do nothing more than attribute to budget issues concerning training. Additionally, and far beyond budget issues, improper or irrelevant training also creates a severe safety problem.

One of the most infamous conditions played upon us is the unremitting reinforcement of an extraneous praxis is the use of the word "survival" to represent any skill, task, or program associated with use of force or situational control training and conditions. Without question, the risk of violence has been associated with the military and law enforcement since their earliest inception. However, it wasn't until tragedies like; the iniquitous Newhall incident occurred, that problems in tactics and weapons handling skills were brought to the forefront of training and administrative circles. It was about this time that the phrase and subsequent profit making industry of "officer survival" was born. As "survival" flourished into the catch phrase used throughout all levels of law enforcement it also exploded into a multi-million dollar industry marketing everything from books and magazines to subliminal tapes and "expert" speakers, all available at outrageous prices.

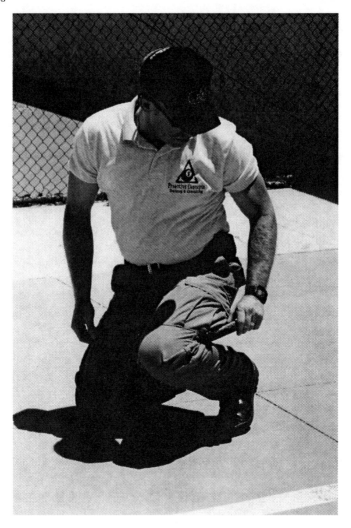

Survival is not an option or even part of the

equation. To win in a high risk environment

we must train for any and all potential

situations, especially the use of the non-

dominant side of the body.

(Courtesy of Perry Taylor)

Survival, by definition, means steps or actions taken after an unintended or involuntary incident occurs, enabling an individual to maintain basic existence until escape is possible or help arrives. During an instructor course that I taught I took the 50 participants through a series of relaxation and visualization exercises. At the point that I felt everyone was relaxed and in his or her "happy place", I introduced the words "survive, survival, and survivor" into their thought process. Following the exercise, I went around the room and asked each participant to describe the first thought that entered their mind when I brought up the topic of survival. Of the fifty different people in the room, not one mentioned anything about going to work, wearing a uniform, or carrying a gun. Oddly enough each person said that they saw visions of sinking cruise ships, Mt. Saint Helens erupting, and air crash victims resorting to cannibalism. All examples of situations where those involved were involuntarily thrust into the event and were totally unprepared to deal with it, suggesting that it is the level of preparation and the voluntary status of the circumstances concerned that separate a survival situation from all others.

Although overworked, poorly paid, and under appreciated, we in law enforcement and the armed forces do commit to our professions on a voluntary basis and we train in order to prepare ourselves for the preordained challenges and dangers of the profession. The better our training, the greater our level of preparation. When something as important as our lives or the lives of others are at stake, should we prepare to accept our occupations as on-going efforts to merely survive? Realistically, if survival was the hinge pin to the equation then wouldn't finding a different line of work be a better solution?

To control and ultimately win in dangerous situations, an individual needs much more than just skills. To win, an edge, both mental and physical, must be developed allowing an individual to successfully resolve a situation regardless of complexity. Potential high-risk situations can take an individual by complete surprise, overwhelming them psychologically and generating a "victim" or "survival" state of mind. To counter this, to distinguish the protector from those who are to be safeguarded, a strong conditioned hunger to win must be tapped. This is commonly called a "winning mindset". To create this mindset, we [instructors] must stop emphasizing survival and start preaching winning. Throughout our lives, from childhood to our adult years, we all have experience with winning and losing, and all but a very few individuals hate to lose. On the other hand, the greatest experience most individuals have with survival comes from history lessons involving shipwrecks, natural disasters, and other such involuntary situations. These history lessons are the earliest and most profound mental images of survival for most people, this is a highly significant factor when dealing with high stress situations, considering the proven power of the mind and its ability to accept suggestion.

Defense or Offense

Regardless of personal sentiments, very few can say that the standard of living in the United States of the late 20th and early 21st centuries isn't 100% a purely voluntary environment. This is especially true concerning the various occupations that require a functional knowledge of all things martial. Granted, from early childhood through the better portion of High School, others generally guide ones path and their decisions are typically steered by laws and policies that mandate that parents send their kids to school and those kids attend minimum required courses. However, by the time most of us reach our junior year in High School we have attained a status that allows us to pick and choose various classes or even drop out if school if desired. Following High school, as we enter our adult lives, we make a series of progressive, often digressive, choices concerning our lives. We voluntarily decide to go to college or go to work, we and we alone decide the course of our lives.

In the world of law enforcement and military circles, we decide on various courses of actions based on the developed objective for a given situation. For example, a lone officer in a desolate location observes a suspicious vehicle pass his position at a high rate of speed. The officer pulls in behind the vehicle and follows it for a period of time to confirm the erratic behavior of the driver and to further develop his probable cause for stopping the vehicle. Eventually, the officer activates his emergency equipment and pulls the vehicle over to the shoulder of the road. The officer notifies his dispatch of the stop and its location. A short period of time later, the officer is advised that the vehicle is stolen and was used earlier in a violent armed robbery in an adjacent jurisdiction. The officer is further advised that the suspect in the armed robbery has been identified and has a number of caution indicators associated with his identity, including a violent propensity towards the police. The officer initiates the procedure for a felony vehicle stop and requests back up. Once the officer's assistance arrives, the suspect is commanded from the vehicle, apprehended without incident, and a subsequent search of the vehicle produces evidence of the recent armed robbery. This scenario sounds fairly typical and somewhat simple, but lets look at the officers actions. The officer observed a possible violation and made the decision to investigate it further. The officer, after developing his probable cause, chose the location for the stop and upon receiving information from his dispatch center he decided on a course of action to take. The entire set of circumstances was a series of decisions that the officer made. Understandably, the decisions were all based off of things presented to the officer, but all the actions were based off of the conscious decisions of the officer. In fact, the career choice that put the officer there was a decision that he made. Now, I'm sure the question comes up as to how this relates to the words offensive and defensive, stand by and I'll explain.

The day-to-day actions, including choice of occupations, are all a series of voluntary decisions, not ongoing efforts to merely survive.

(Courtesy of Perry Taylor)

For years we have been told and re-told how police engage in defensive operations. During the academy, cadets learn defensive tactics and learn to defensively drive a vehicle. In the military, the Military Police, Rangers,

13

Special Forces, Seals and others are taught unarmed self-defense. But are these techniques or the thought process behind them really defensive? If you have voluntarily decided to apprehend someone, you have made a decision to move and act in an offensive mode. It is your goal to seize another human being, to restrict that individual and to remove his civil rights. You are going forth into this situation with the purpose of taking someone down and are voluntarily committing to follow through with the act, regardless of the suspect's subsequent actions. These voluntary actions put us in a position where we are not responding; in fact we are initiating the action against another. This is a purely aggressive and offensive action. In a defensive situation we would have ventured into it involuntarily and without any fore-warning of potential danger. It is my contention that many situations are, from start to finish, offensive in both form and concept. However, we have been told that all we do, be it civilian or military, are defensive. Further, we have been conditioned to apply deadly force and had the word "defense" attached to it. Perhaps this is an industry attempt to limit liability or to reduce the potential of offending someone, but there is no getting around it, these various activities are offensive. The problem arises in the initial training of the individual when he or she is told that they are to act defensively and they mentally align this conditioning with a preconceived designation of the term defense. Later in training we tell this individual that they must further develop a "winning mindset", but how can one win when one must continuously defend? In order to win, at some point in time, you must go on the offensive.

In order to dominate and control a situation the individual must take actions

that are offensive in nature.

(Courtesy of SG Five)

The Winning Mind

Without a doubt one of the most overlooked areas of training that probably comes with the greatest amount of limitations is that of the development of a winning mindset. All too often this is called the "survival mindset". It is important to remember and to express that in the middle of a violent situation is not a survival atmosphere. Survival is the culmination of actions taken after an incident has occurred, like the cannibalism of the dead so the living may go on. In the area of emergency services, survival refers to those actions taken to recover from an incident, either physically or emotionally. Survival is what we do to live healthy for years after an incident occurs. Instructors must take a more proactive role in encouraging their students to believe in themselves and their abilities, not a false sense of security, but an actual belief in the effectiveness of their actions.

The winning mindset is a state of mind that puts an individual into a mode that allows them to focus on the task or situation at hand. It is a presence that is generally void of emotion, where awareness, analysis, and response become a fluid process. Although there are many mental variables associated with a real winning mindset, the one constant is the individual's confidence in their skills, life values, personal beliefs and a belief in the mission at hand. These elements help to create a mindset that allows the individual to focus on the mission without distraction and to respond without hesitation. Instructors have a direct influence over all of these elements except life values and personal beliefs, these two components are products of an individual's life experiences and are probably not effected by the short duration of most training programs. However, I can state with all confidence that values and beliefs play a significant role in cardiac and adrenal activity during high-risk situations. So long as instructors understand the relationship theses elements have in the development of the

Wes Doss

mindset, they should be able to adequately express it to students, so that the

student can deal with these issues privately and personally.

Chapter Two

"Learning is not a spectator sport"

-D. Blocher

Being Human

The study and application of combative skills is as old as human existence itself and have long been a means of establishing dominance or hierarchy within a group. In fact the practice is still used today throughout many species in the animal kingdom and even some of the more remote human civilizations. One only needs to spend an afternoon watching one of the numerous educational channels to see glimpses of wild horses, bighorn sheep or primitive bushmen violently battling for dominance over their brood. However, in today's modern world of political correctness the study of conflict has become frowned upon and is generally considered an inappropriate form of behavior. This behavior is often considered the playground of the more dysfunctional or sadistically oriented members of our society and is generally only considered acceptable conduct or behavior for those in law enforcement, military and similar protective occupations.

Oddly enough, while society professes to hold a negative view of combative themes, it seems to maintain an unprecedented affection to its visual and fantasy oriented side. In fact, the visual appeal or fantasy is very evident and seems to be an almost intrinsic trait of mankind. Humans find fascination and arousal in the emotional stimulation and flamboyant moves that are found in the movies, popular computer games, "reality-based" TV shows and even popular talk shows. From childhood through our senior adult lives, we witness people being thrown through windows and knocked through the air with a single punch or kick. We see heroes and villains flying through the air firing weapons and obtaining phenomenal hits and causing instantaneous incapacitation. Further, we have witnessed mere humans surviving tremendous kicks and strikes to the head with no trauma, enduring fistfights that last longer than any professional boxing match in history. This runs right along with the images from the 1970's cop shows where the opponents could fire hundreds of rounds from five shot snub nosed revolvers never needing to reload and could emerge from the water with perfectly styled dry hair.

Many trainers and training organizations have fallen prey to the flash and display of the unbelievable scenes we find throughout our world, teaching techniques and selling clothing and equipment that provide more style than function. Many well known trainers have advanced their curriculum to include extravagant, unnecessary movements that are not practical in any way, shape or form.

Having fallen prey to the flash and display

Of many martial arts styles, some trainers

have forgotten the purpose of their trade and

advocate techniques that are often

impractical.

(Courtesy of the author)

It is important to understand and to never forget, that any true form of combat is a very real, calculating and often very physical activity. It is then reasonable to assume that an individual's physical and mental capabilities will have tremendous weight on the outcome of the situation. To make a positive impact on both the mental and physical aspects of combat the training must be designed to effectively tap both of these mechanisms. Further, the trainer must have, at a minimum, a functional knowledge of how humans learn.

Why Most Training Fails

Most organizations, unfortunately, establish training around the money that is available. Therefore, it is important for these organizations to use their training dollars as effectively as possible. However, concerns of customer satisfaction and cost containment are useless notions for these organizations to consider because in training you typically get what you pay for, whether you realize it or not.

One of the primary causes for wasted training dollars is ineffective methods. Too often, organizations rely on lectures, speeches, videos and unqualified "experts" to provide their training, falling back on the idea that it must be working because no one has been hurt yet and we have not been sued. While these methods often get high marks from the participants, research (typically ignored by many training "professionals") shows they rarely change behavior.

Watching and knowing is far from doing; good intentions are far to easily laid to rest by old habits. Another way of wasting training time and money is failing to link training with the actual needs of the students and to the day-to-day operations of the organization. Vividly displaying why what happens in the classroom and what happens in the real world are truly worlds, if not universes, apart. Training that produces tangible results starts by changing behavior. Changing behavior will change attitude and ultimately effect an individual's confidence and confidence is essential for winning in a terminal situation. Most administrators and managers (folks who should know better) typically get this backwards.

Training that produces tangible results starts

By changing behavior and deeply effects the

student confidence.

(Courtesy of SG Five)

Wes Doss

How Adults Learn

A tremendous part of being an effective instructor involves understanding the various methods of how adults learn and designing training programs that apply these methods best. Instructors who understand how humans learn and how they can help their students reach their potential have an understanding of how the brain functions. All healthy adult humans have a literal inexhaustible capacity for learning. In comparing adults to children or teens, we find that they have some very specific needs and requirements. One of the most profound pioneers in the field of adult education was Malcolm Knowles. In his research, Knowles coined the term andragogy as the art and science of teaching adult humans and focused more on the process of learning and less on the content being taught. In an environment of andragogy, instructors perform more as facilitators or resources and less like lecturers and evaluators. In his research, Knowles characterized five assumptions about adult learning.

1. Self-Concept: Or autonomy is where a person matures from a self-concept of being dependent towards being self-directed. This allows the students, while under guidance, to discover things for themselves and actively involves them in the execution and evaluation of the instruction.

2. Experience: This is the ever-growing pool or reservoir of experience that an individual accumulates throughout life and becomes a tremendous resource. All experiences are important, especially mistakes.

3. Readiness to Learn: Or goal oriented towards what the individual wants or expects to attain from the training or from their position in society.

4. An Orientation to Learning: As an individual matures, his or her time perspective changes from deferred relevance of knowledge to one of immediate application. As an individual's orientation towards learning changes, his or her perspective shifts from one of subject focus to a perspective more aligned with problem focus.

5. Motivation to Learn: Knowles presumed that as an individual matured, their internal motivation to learn would also grow (Knowles 1984).

As with many other elements of modern society, and education especially, these assumptions and claims made by Knowles have been the subject of considerable debate. It is important to understand that Knowles' concept of andragogy was an attempt to develop a comprehensive theory about adult learning that was based on the actual characteristics of adult learners. Additionally, Knowles made extensive use of a model of relationships derived from humanistic clinical psychology and the qualities of good facilitation. Knowles was also concerned with curriculum development and behavior modification, encouraging both the trainer and the learner to identify needs and set objectives.

Unlike children, adults typically carry around a sizeable quantity of responsibilities; like time, money, childcare or personal relationships, that they have to balance against the demands of learning; this is critical for trainers to keep in mind. Although generally devoid of such distractions, even trainees in military basic training enter the training environment with some quantity of emotional or moral baggage. These responsibilities create barriers against participating in training and successfully assimilating the information, thus causing motivational factors.

Motivation problems are probably the most pronounced barrier to training. What motivates adults to learn? Things like competency, promotion, licensing requirements or the need to learn in order to comply with company or organization directives can be considered motivators. However, as use of force instructors and dealing with topics that involve the taking or maintaining of human life, we need to understand how to motivate people to apply force and to break down the barriers that these individuals will have to training, and more importantly the barriers to assimilating the information provided during training. In this process, it important to determine why the student is attending the training (initial motivator). Once this is known, the instructor needs to plan a strategy of motivation that continuously shows the link between the training and the expected or potential outcome from the training.

If you continue to research the field of adult education and training or even continue to read on throughout this book, you will find that motivation is a pivotal concept in most theories of learning. Motivation is closely related to arousal, attention and anxiety. A particularly relevant point on motivation is found in the concept of achievement motivation. Motivation to achieve is a primary function of an individual's own desire for success, the expectation of success and the various awards that can be gleaned from it. This manner of motivation is primarily associated with career goals and school grades, but can be adapted to our field of instruction. Many of those who we deal with, and I know this sounds strange, but are not really aware why they have to learn to fight, shoot or control a situation. The fact that many of today's young police officers will go a full 20 years without ever getting in a fight is one of the reasons.

Learning and Learners

Use of Force training, as a form of adult education, is an incredibly complex and confusing process that is grossly misunderstood by many who are involved in it. This is not necessarily their fault, as most are simply providing the type of training that they are accustomed to and received themselves, maintaining a consistent but ineffective state of perpetual motion, like a cow drinking milk. Others are teaching a curriculum that is mandated by their agency or by some sort of governing board, often resulting in a "square peg, round hole" inflexible and generic result. This lack of understanding of the training process can create a hamper to the educational development of the student and craft a serious safety problem when the student is faced with the realities of their chosen profession.

Good instructors should strive to break the"square peg, round hole" method of

training and help students understand the realities of their profession.

(Courtesy of SG Five)

When contemplating the process of training it is important to understand

that adult training or education, much like law and psychology, is a science of theory, assumption and speculation. The world of adult education is most definitely a world of many schools of thought, with each school subscribing to the theory or assumptions of one of the many well-known figures within the discipline. I have already mentioned Malcolm Knowles and I acknowledge that Knowles is one of the most significant pioneers in the field. However, please understand that this field, especially the arena of use of force training, is relatively new as a science or system. In understanding that the field of adult education is a new discipline it is important to consider all available theories and concepts and reflect on how they might make your individual efforts stronger.

When examining other concepts about education and most importantly about human behavior, it is inevitable that one will come across another very prominent character of this field, Edwin Guthrie. Guthrie became one of the most significant figures in the field of behaviorism. His studies examined the concepts of association and its limits in attempting to explain how learning takes place in an individual organism. Guthrie hypothesized that there is only one type of learning and that all principles applying to one instance, apply for learning in all instances. Further, Guthrie assessed that differences in learning do not evolve from there being different kinds of learning, but rather from different kinds of situations.

Edwin Guthrie's most significant development, at least in our field, is probably his Contiguity Theory. This theory, having found tremendous validation for years, states that all learning is a consequence of association between a particular stimuli and a particular response. In addition, Guthrie maintained that stimuli and responses affect specific sensory-motor patterns, assuming that what is learned are movements and not just behaviors.

The Contiguity Theory further implies that failing to remember a skill or task is due to interference as opposed to a lapse of time. This simply suggests that stimuli becomes associated with new conditioned responses and that previous training can be changed by being associated with inhibiting responses like fear, depression or fatigue. Putting motivation in a role of creating a state of arousal and activity in order to generate responses that can be conditioned.

This concept is classed solely as a theory because while Guthrie did apply this work to humans the majority of his experimentation was done with animals. In fact, the classic paradigm if the Contiguity Theory was done with cats learning to escape from a puzzle box (Guthrie & Horton, 1946). A glass box that allowed the cats to be photographed was used in the experiment. The photographs were taken and subsequently used to show the repetitive sequence of movements done by the cats as they learned to escape from the box. This experiment established a number of principles associated with training and behavior modification.

1. In order for behavior conditioning to take place, the organism or student must earnestly participate and respond in the training. In other words, the student must be doing something.

2. As the learning involves the conditioning of specific movements, instructions must be given with every specific task.

3. The student must be exposed to several variations of the stimulus in order to produce a basic desired response.

4. The student must do, or at least believe he has done, a correct response at the completion of the training. This is because we want the student to leave training with a positive impression of the training and the final response is that which the student will most likely associate with in the future.

Not in total opposition to Guthrie's theories but clouded in just enough conflict is the Schema Theory developed by Richard A. Schmidt in the course of his work in the field of motor learning or motor skills learning. Schmidt defined motor learning as; a set of internal processes associated with practice or experience leading to relatively permanent changes in the capability for responding (Schmidt, 1988). This means that individuals do not learn specific movements. Instead, they construct generalized motor programs, learning the relationship between the parameters of an action and the outcome. An example of this would be an individual practicing a movement, like throwing a ball at various distances or directions or climbing stairs of various dimensions. As the movement is practiced, information is collected internally and a greater understanding of the correlation between the movement, the outcome and the individuals control of the parameters of the situation. Schmidt theorized that an individual will learn more quickly the relationship between manipulating parameters and achieving a desired outcome if they practice the task in a wide variety of situations and experience mistakes in the process. In his research, Schmidt determined that the numerous relationships between the parameters of a situation and the potential outcomes are subconsciously collected by the individual and stored in two "schemes" or "schemas".

1. Recall schema: Relating outcomes to parameters like the duration of a movement and the overall force required to produce or execute a movement.

2. Recognition schema: Relating to anticipated sensory consequences of a movement to the movement's outcome. Creating an internal reference of correctness (Adams, 1971)

For clarification purposes, I will try and explain the concept of schema. F.C. Bartlett (1932, 1958) is the primary researcher credited with developing the concept of schema or schemata. Bartlett developed the concept based on his studies of the human memory and the ability to recall details from stories where the details had not actually been provided. In one experiment, test subjects were shown pictures and photographs and then asked questions depicting the theme of the picture. Each person would remember different details about what they saw. Bartlett suggested that our memory takes the form of schema and provides assumptions within a mental framework to help us remember and understand information. Now, is that crystal clear? I doubt it. Essentially, the idea behind schema is that our brain, primarily our memory, inserts assumptions about a situation based on our prior experiences and training. Because schemas are acquired over a lifetime of learning, they are largely responsible for guiding us through our perception and problem solving situations, functioning like a screenwriter crafting a movie script.

Helping students understand tasks and building experiences

helps develop the ability to problem solve in challenging situations.

(Courtesy of Perry Taylor)

The mind develops a running history of

experiences or schema. The schema

provides a fill for gaps in the perception of a

situation. The best experiences come from

quality, hands on training.

(Courtesy of the author)

Cognitive Learning

As we explore the education and psychology pioneers that I have mentioned in this text, and their various theories concerning how humans learn, it is important to discuss the diverse dominion of cognitive psychology and the learning models that have been developed from this discipline. Cognitive styles technically cover the different ways that we humans process information, highlighting how an individual thinks, remembers and solves problems. Where other practices try to identify varying abilities, cognitive systems simply designate tendencies to perform in a specific manner. Cognitive styles are often characterized as various personality capacities that influence our attitudes, values and our social skills.

One of the most well defined concepts of cognitive learning is known as the Cognitive Load Theory (Sweller, 1988). This particular theory suggests that learning occurs best under conditions that are aligned with the cognitive architecture of the learner. In this architecture, the short-term memory is a relatively small device, limited in the amount of programs it can run simultaneously. It is believed, in this theory, that a combination of various elements (schema) is the cognitive structures that create an individual's base of knowledge.

Long-term memory is somewhat more sophisticated and more multi-task oriented than the short-term memory, allowing us to perceive, think and solve problems. Again, schemas are what permit us to deal with multiple elements as a single element. Once again, these are the structures that make up our base of knowledge. Understandably, different individuals have different bases of knowledge to draw from, separating the experienced operator from the novice. This separation exists only because the novice has not yet acquired the experiences or developed the schemas of the more worldly counterpart.

Learning occurs when a change in the existing schemas of the long-term memory takes place and the change is verified through performance, showing a progression from clumsy, mistake-prone, protracted and arduous to smooth, fluid and graceful. The material is learned and the change occurs because the learner becomes progressively more familiar with the material. The cognitive uniqueness associated with the material are changed so that it can be digested more efficiently by the working memory.

From an instructional perspective, information contained in instructional data must first be assimilated by the working memory. For schema procurement to take place, the instruction should be designed to reduce working memory burden. The cognitive load theory is based on developing techniques that reduce the load to the working memory in order to facilitate the changes in long-term memory associated with the acquisition of schema.

Operant Conditioning

Among the many theories of adult learning, probably the most relevant and widespread is that of operant conditioning. This hypothesis involves increasing a specific behavior by providing a reward or reinforcement for performing correctly, or limiting the behavior by following the undesirable act with punishment or negative feedback.

One of the most notable figures in the field of adult education is behaviorist B.F. Skinner. Skinner construed that learning is essentially a change in overt behavior and that changes in one's behavior came about as a result of an individual rejoinder to particular stimuli. A response to an action generates an end result or product; like hitting a ball, throwing a punch or pulling a trigger. When an individual in training is faced with a particular stimulus-response pattern and receives motivating reinforcement, that individual becomes conditioned to respond.

Reinforcement, as in many theories, is the operative element in this concept. Anything that strengthens a desired response is generally considered reinforcing, particularly some form of positive feedback or criticism. Positive can be in the form of verbal praise, a feeling of accomplishment or self-satisfaction. As previously stated, and it should be fairly obvious at this point, that each figure in this field had a different take on how we learn. Does this varying difference in theories mean that one is any more right than another? No, certainly not. In fact, although differences exist, there are a number of consistent themes that each of these presumptions shares. This is what we in the cop business call a clue.

1. Optimal learning occurs when relevant information is presented and the learner understands the value of learning it

2. Optimal learning occurs when the learner is involved at a "hands on" level and actually has to perform the skill.

3. Learners need to be challenged under conditions that accurately replicate actual conditions where the skill is expected to be performed.

Additionally, there are two other themes that run consistently through all theories of adult education, motivation and feedback.

Motivation and Feedback/Reinforcement

One of the most pivotal concepts in learning is that of motivation. In fact motivation is closely tied to many other human factors, including the sympathetic nervous system activation that effect learning and stress mitigation. In most cases of behavioral science, motivation is considered to strictly be an element of our primary human drives such as hunger, sex, sleep or comfort. In terms of learning motivation, it is widely believed that learning itself reduces our drive and therefore a substantial amount of external motivation is required for successful learning to occur (Hull, 1943). In his study of how variables like drive, incentives, inhibitors, prior training and prior experiences affected behavior, C. Hull determined that motivation and subsequent reinforcement were decisive attributes that contributed heavily to the learning process. One of the most important concepts in Hull's work was what is known as the habit strength hierarchy. This concept related that an individual can react in a number of ways to a given stimulus and the likelihood of cultivating a specific can be altered by positive training, but can also be affected by other variables (e.g. fears, inhibitions, experiences). Essentially, this means that the degree of learning can be manipulated by the strength of a particular drive and its underlying motivation.

Still others in the field of cognitive thought saw motivation as an achievement function based on an individual's personal craving for success and the expectation of success (Ames and Ames, 1989). Various studies have shown that generally humans prefer tasks of a medium level of difficulty and that particular individuals, with a deep yearning to achieve, obtain better performance in programs that they perceive as important and highly relevant to their jobs.

Feedback and its sister concept of reinforcement are constituents that work in unison with motivation. Generally, feedback is externally received information about an individual's response or performance. Feedback can be positive, negative or neutral. Reinforcement on the other hand is generally geared to influence a specific response or circumstance and is either positive or negative with very little middle ground. Reinforcement can be from both internal and external sources.

Most learning theories place substantial emphasis on motivation and feedback/reinforcement since an understanding of what we are doing is essential for identifying and correcting mistakes, developing new skills and motivating us to learn. There is, however a very crucial variable associated with these concepts and that is time. By time, I mean the appropriate timing to apply motivation, so that it appears sincere and not patronage, and the length of time between an event and the application of feedback/reinforcement, so that the relevance of the feedback can be identified.

Feedback and reinforcement are integral elements to a students experiences.

It's crucial for an individual to know what they did right or wrong and how to fix it.

(Courtesy of Perry Taylor)

But How Do We Motivate? Especially Those Who Are Unmotivated.

I have seen it time and again, teachers and instructors literally beating their heads against the wall over students who seemed to refuse to learn. I have been part of lengthy conversations filled with sighs and grunts of frustration and colorful adjectives concerning unmotivated and unwilling students who seem to hate being in the class, I think most teachers and instructors probably have. However, I have come to regret many of the four letter "pet names" I've used and I have learned to spend more time understanding my students, their needs and their expectations. In all reality, the students themselves harbor much frustration over training and have a difficult time shaking these annoyances during the brief periods that we actually deal with them.

Most of the adult students today are somewhat older than the stereotypical post-teen that most of us envision in school. For those of us teaching use of force to military, law enforcement and security professionals, we are likely to encounter students who are somewhere between their mid-twenties and their late fifties. These older, more seasoned individuals often bring a tremendous amount of life experience and career involvement into the training environment, causing these folks to be just a little apprehensive about being taught. Additionally, adult learners,

43

especially older adult learners, often lack real study skills and usually have other things on their minds like; work issues, family problems and financial concerns. If all this wasn't enough, many actually view a training environment as a challenge to their social identity and a major hit to their confidence. Many come from significant or diverse backgrounds and have been shaped by their various experiences in a far different way than the material that is being presented to them, causing many to don the attitude of, "you can't teach me anything that I don't already know". When you consider the social and psychological anxieties that many of these folks have, it really is no wonder why they don't always understand the importance of the training.

Its important to always remember that all students are different and bring different attitudes to a class. In order for everyone to embrace the material equally the instructor must accept these differences and work to change attitudes.

(Courtesy of SG Five)

All instructors, especially those who develop their own programs and curriculum, like to believe that their material will be something that is cherished and admired by their students. Further, we all want classes filled with active participants who crave assistance and insight, thriving on the challenges presented during our program. However, this is not very realistic and we end up often blaming the students for it not being this way, seeing them as passive, uninvolved lemmings.

How do we reach those who seem unreachable? How do we make this change? First, we take responsibility for the productivity of our classes and refuse to blame the student. As odd as this sounds, many individuals have been conditioned to sit down, shut up, listen and learn, not being aware that they have the right to get involved and speak up. This production is nothing more than an attempt on someone's part to perpetuate the axiom that learning is some impersonal process where you either make it or you don't and if you sit and memorize information, you'll get just enough to pass a test. Well, this may work in some college classroom environments, but in this field where the real test is when the student is nose to nose with a dedicated opponent, that will not do.

Teachers and instructors who can create warm and accepting, yet professional atmospheres will promote enduring effort on the part of their students and this leads to more favorable attitudes about learning. Letting students feel like they are appreciated and more than a face in the crowd helps to promote an atmosphere where the student wants to be there. Additionally, recognizing individuals for their talents and abilities goes a long way to developing rapport with your class, especially when this recognition is sincere.

An instructor needs to be perceived as a "real" person and not someone who demands to be placed on a pedestal. Far too often, instructors see themselves as a deity or demigod that should be revered and is always looking down on those in his class. Instructors in this category often devise assessment or testing techniques that aren't designed to challenge the student, but rather embarrass and decimate any confidence the student had. To motivate and support subsequent motivation, instructors need to monitor their students and support their learning process, informing them of their progress and challenge them at a level that is appropriate for the level of training. Realistically, learning, particularly in the arena of force application, is the learning of critical new skills and ideas. If the student is unaware of what he or she is doing wrong or how he or she can do it better then what, exactly, are they expected to get from the training? Worse yet, what is the potential end result from this type of "training"? One or two good guys die?

This represents an unsatisfactory situation! A situation that none of us should ever settle for, but frequently do.

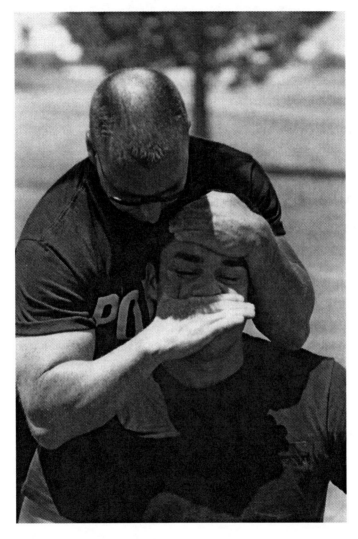

All too often instructors allow their egos to become too great and the student ends

up paying the price.

(Courtesy of Perry Taylor)

A committed and concerned instructor takes

the time to ensure that the material is understood.

Courtesy of Perry Taylor

The use of "live" blades in a training class is a dangerous proposition. Our goal is to train for realism, but not at the expense of our students.

(Courtesy of Perry Taylor)

Uncontrolled or unsupervised actions can

end in tragedy. Students need to respect the

potential outcome from any technique or

tool they choose to use, but at the same time

be trained to use it under appropriate circumstances

(Courtesy of Perry Taylor)

As you can surmise so far, human education is an incredibly profound field, far deeper than most imagine and is continuously growing. The problematical world of teaching adults gets even more complex for those of us in the arena of martial training, as we must not only deal with the same issues as a classroom lecturer would, but we must also deal with overcoming the moral, psychological and even physical problems associated with our craft. Because the environments where our students will apply their skills lack the space and time requirements for checking information and asking for assistance, we must condition our apprentices to react both quickly and appropriately.

The nature of our job and our student's jobs puts us in environments with limited space, distance, and time. Students must be conditioned to respond quickly and appropriately.

(Courtesy of the author)

The human brain is an extremely complex mechanism, one whose complete potential is still not known. In simplistic terms, the brain the associated nervous system can be considered an intricate wiring harness and attached computer module. From a point in time, occurring before birth, our bodies begin to run this wiring creating various paths of fibers that establish communication and electrical connections from neurons to neurons, establishing both these fibers and new neurons at a rate of nearly 1000-2000 every second until the nervous system is fully developed with nearly 15 billion nerve cells created. Once these neurons are established, each function like a network computer terminal connected to the main server brain. As we mature and gain experiences, our network of nerve cells react to the various stimuli and experiences in complex and specific patterns. Each time the stimulus is reintroduced, the association of these network terminals grows stronger. This is the basis, in very laymen terminology, for establishing and developing motor skills. As a technique applied to a motor skill is repeated and greater proficiency is gained, the established pattern gets stronger (Siddle, 1995). Now these examples are very simplistic, as many other variables come into play in developing motor skills, including chemical and protein reactions in the body and greater refinement in the overall neural process. These other variables are complex and can be somewhat confusing. What is important is to acknowledge that the development of motor skills is a critical component of what we do. In fact, motor skills are an extremely important element in learning and teaching a significant portion of what we do in the use of force field. Motor skills are generally classified as continuous, discrete, or procedural movements. The last category is the most relevant to real world applications like driving, fighting and firing a weapon. Long-term retention of a motor skill is generally considered to be highly dependent on regular practice as repetition after a task is learned and refresher training typically show that the student will have a lower likelihood of forgetting the task. However, before we can consider repetition or refresher training, we have to make sure that our students obtain and absorb the information initially and that we correct all undesirable motor behavior, while effectively prompting and guiding our students. Generally, there are two ways to facilitate this in the realm of motor skill development: (1) slow down the rate at which the material is being presented, and (2) reduce the amount of information that needs to be processed (Marteniuk, 1976). Several forms of motor skills or motor behaviors are learned through imitation, this is particularly true with complex movements like those found in many forms of defensive tactics, martial arts or even firearms training. This is a simple concept, we must crawl before we walk, walk before we run and run before we win.

In the process of motor skill development, it is tremendously important to guide the student, and when necessary, prompt the student even when we are trying to let the student work a problem through trial and error (Singer,

1975). Generally, some form of guided learning is required when a high proficiency is required in the learning of a new skill. Further, when the task learned is to be recalled and conveyed to new situations some type of problem solving strategy is needed. This could be the case where basic marksmanship, learned on a static range at set distances, is applied inside a building or from the cab of a patrol car.

There is also significant evidence that suggests that mental preparation, the use of imagery, can facilitate some very sizeable gains in performance. This is probably because the imagery, done at any time, allows the assimilation of supplementary memories of interrelated physical tasks (schema). It's important to note that the body and the mind really don't differentiate between reality and mental images. A man who dreams of falling often wakes up suddenly in a pool of his own sweat and with a racing heart rate. To both the mind and the body, the dream was as real and as threatening as reality.

Chapter Three

"I think, therefore I am"

-Descartes

The SNS & PSNS

In all general appearance, use of force training appears very simplistic. A bunch of guys at the range or on the mats, having a good time, a little male (or female) bonding. However, the truth is, as demonstrated by the complexities of human learning in the previous chapter, that it is a truly complex situation. Combine this complexity with the inherent limitation of the human brain to mitigate stress and the problem gets even worse.

The little gremlin within all of us, the thing that really is what makes or breaks us in stressful situations is the human nervous system, or the autonomic nervous system. The nervous system is an extremely complex network of communication lines that run all throughout our bodies and play a part in controlling every function that our bodies have. The human nervous system, for the purposes of this manual, can be broken down or divided into two distinct and separate sub-systems. These sub-systems are the para-sympathetic and the sympathetic nervous systems. In the average, day-to-day, low stress world that most of us reside, the para-sympathetic system is dominant. While functioning under the control of the para-sympathetic

56

system, healthy humans are in full control of their bodies and maintain the ability to speak clearly and employ complex or fine motor skills. This is to say we are normal, able to function and interface with our fellow members of society all day long and in a variety of situations. However, when a human is confronted with an unprompted and unanticipated event, especially a threat, the para-sympathetic system goes on vacation and the sympathetic nervous system, or SNS, moves in.

Due to the various complications created by

our own bodies we react very unpredictably when involved in a stressful situation,

even more so when we are injured. Making the training of our non-dominant limbs

important.

(Courtesy of Perry Taylor)

When the SNS is activated, the heart rate increases, even more so when we are injured. Making the and blood pressure increase stridently and blood is diverted away from the digestive system and small muscles to the larger muscle groups of the body. The increased arterial pressure, rapid heart rate and shift in blood flow cause some marked changes in the bodies gross motor control and physical strength, increasing both sharply. However, this change also causes a severe deterioration in the body's fine and complex motor skills. In other words, when faced with a spur-of-the-moment menacing situation, we as a species get stronger and faster, but lose the ability to do detailed things, like; write our name, drive a car and in some situations, fire a weapon.

When the SNS is activated, it is done completely involuntarily; it is an automatic and literally an uncontrollable response. The activation of the SNS is considered a reflective action and it is based on the perception of a threat or a fear, remaining in control of our bodies until the perceived threat is perceived as gone or eliminated. The SNS sounds like a horrible thing and in many instances it is. However, it is important to remember that it is a prime component of the bodies "Fight or Flight response". This response is often illustrated in the story of the frail little old lady who is able to lift a car off a trapped child. It is in this response that we find some of our most significant and often fatal performance limitations.

Throughout each day our brain receives continuous intelligence about our environment through our five senses. When the SNS activates, the brain constricts our perception to the most dominant and the most reliable of the senses, our sight. This reduction in the ability of our senses includes the limiting of our auditory sense, sometimes to the point of actually reaching deafness. As the eyesight becomes the primary source of information the eyes begin to go through some significant changes. The pupils dilate and blood flow to the periphery of the retina is greatly reduced causing the effective use of monocular vision, or one eye, to be greatly diminished. Monocular vision or the use of one eye at a time is the primary way most individuals are taught to apply basic marksmanship fundamentals. The body resists closing one eye at a time and will actually force both eyes open in an attempt to keep an already narrowing field from closing completely. This decreased blood flow to the periphery of the retina is typically what is termed as "Tunnel Vision" and can narrow the available field of vision by as much as 70% (Breedlove, 1995). As the peripheral area of the retina closes in, the pupil starts to dilate. This dilation causes a marked loss in near vision and an acute interference with the eyes ability to focus, which also contributes to a loss of depth perception (Cannon, 1915).

The eyes are not the only component to endure severe changes; the heart itself begins to drive too hard and causes extreme changes in all effected organs and systems in the body. This stress on the cardiovascular systems generates greater problems. At our resting heart rate, which varies somewhat from person to person, we are capable of performing all things that we have been conditioned to do. As our heart rate increases, our abilities generally decrease, specifically our more multifarious motor skills. As the heart rate reaches about 115 BPM, our ability to perform fine motor skills begins to deteriorate distinctly. This level of deterioration is even noticeable in the performance of highly conditioned endurance athletes who train based on their heart rate. When the heart rate escalates beyond 145 BPM, both fine and complex skills, especially eye-hand coordinated skills, degrades rapidly. At about 175 BPM and higher, the thought process itself begins to weaken, giving way to the onset of hypervigilance and irrational behavior (Siddle, 1995). However, our problems do not solely revolve around an elevated heart rate, as several experts in the field have conducted experiments where they took monitored doses of heart rate elevating compounds, like epinephrine and were able to reliably and accurately manipulate a weapon or navigate a course of fire. The problem really begins when stress from a perceived threatening situation gets so high, over and above an individuals training and experience level, that their performance declines. One of the earliest principles developed to explain the relationship between arousal and performance was "The Inverted-U Hypothesis", the basis of the Yerkes-Dodson law (Hebb, 1955). This supposition states that as arousal or stress becomes too great, performance would deteriorate. The principle was visually expressed on a diagram of an upside down or inverted U showing the increase of stress and a comparable increase in performance until the apex of the inverted U is reached; at which time, performance declines as stress continues to escalate. In simplistic terms, it means that we need some stress in order to perform optimally and so long as an individual still feels capable and confident they should be able to perform with relative efficiency, but when they reach a point where they begin to doubt their ability to manage the situation their performance, starting with fine motor skills and rational thought, would start to decline rapidly.

Exhaustion and confusion are typical after effects from being involved in a high

stress situation, even for highly trained individuals.

(Courtesy of the author)

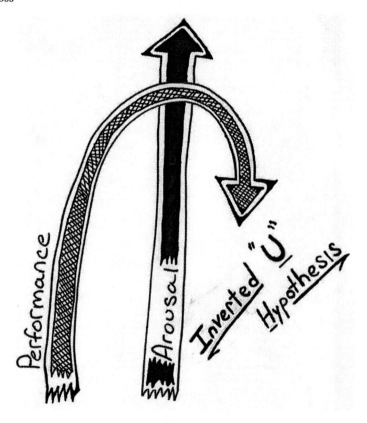

Increases in arousal are paralleled by increases in performance, but only to a point.

As arousal becomes too great the performance begins to decline.

(Illustration courtesy of the author)

It is a well-known fact that when humans perceive and respond to stress, the body increases the production of various compounds and substances and releases them into the blood stream. These elements cause various things to happen and change including an increase of blood to the extremities and an increase in an individual's strength, explaining why gross motor skills like running, pushing or pulling can be performed optimally under high levels of stress. These very same elements that can turn little old ladies into super heroes also makes fine motor skills and cognitive thought inoperable. These issues suggest that it is probably wiser to limit the complexity of training and try to keep as close to the realm of gross motor skills as we can, since under duress fine and complex skills are difficult to impossible. This is not to say that I believe that the development of fine and complex skills should be abandoned, quite the contrary. Since the activation and severity of the SNS is based on nothing more than the perception of a situation, it seems reasonable to believe that if fear and risk perception can be mitigated in training then the use of fine and complex skills may be capable under various high risk situations.

Fears

Our memory is a faculty of our psychology that represents the capacity to carry elements of an experience across time. The memory is truly a remarkable asset, finding itself in nearly every biological process in our environment. In fact, it has been documented that through a very complex process, elements of experiences can be carried through to new generations of a species. Throughout all the biological entities on our planet none have the advanced capacity to make and store interpretations and images of the external world as the human brain. This storage function involves an extremely complex neuromolecular process of intercellular and synaptic architecture and is far outside the scope of this material. However, it is simple to say that the brain and the memory change with each and every experience, good or bad.

The brain, through the various senses of the body, allows an individual to perceive the environment, process the information, assimilate it and store it. These perceptions are the basis for what is left of our human survival instinct and allow us to promote our well-being and optimize our chances of successful procreation of our species. In order to do this, the brain creates internal depictions of the external world, or the world outside of the mind. In creating these depictions, the brain takes in information and transforms it into neuronal activity and creates an as needed library of these representations (Kandel and Schwartz, 1982). Further, the brain creates associations between sensory information (sights, sounds, smells and emotions) and specific events, allowing us to individualize our stored sensory information and use it in future events. This can be characterized by a small child touching a hot stove and feeling the pain from the intense heat. As the child matures he sees a stove and realizes that it can be hot and that he may need to exercise caution. However, it is important to understand that the mind and the body cannot differentiate between real or imagined events.

Some events witnessed on TV, movie screen, or in dreams can have a significant impact on actions in real life. In the 1970's, the ground breaking movie "Jaws" was released and following the public's exposure to it there was near pandemonium. Droves of people would run screaming from all bodies of water at the mere mention of a shark, this included inland fresh water waterways and swimming pools. Very few of the people had actually witnessed a shark and even fewer had ever been attacked by a shark, but the captured image stored in the inner sanctum of the mind was enough to create a very real fear.

The topic of fear, although certainly debatable, is an ancient one. Thucydides considered it to be one of three strongest motives for human action, followed by honor and interest. Aristotle concluded that an individual's response to fear was a solid indicator of the individual's capacity for ethical behavior. Regardless of one's beliefs about fears, it is important to realize that they represent a real issue when you engage in martial activities and represent a greater issue when you are responsible for the training of those who function in pugnacious lifestyles.

Many believe that they are born with fears, however very few newborn babies enter into this world with fear, instead they learn to be afraid. Fears or phobias are learned as a product of a rational emotional response to an individual's perception. Simple phobias are fears of specific things or situations like, heights, confined spaces, animals, and the dark, simple phobias sit in marked contrast of agoraphobias, which represent the fear of places or social situations. Both types of phobias are learned or conditioned behaviors that afflict about $1/10^{th}$ of society and are usually not suffered until late adolescence. Regardless of their appearance, fears are learned. In the 1920's an American Psychologist by the name of John B. Watson and his assistant Rosalie Rayner conditioned a small baby to have a deadly fear of a small white mouse by associating the sight of the mouse with a disturbingly loud noise. Now by today's standards, this experiment would be closely scrutinized and probably not allowed. However, the experiment did show that humans can learn to fear harmless things when they are associated with unpleasant stimuli. This is vital to understand since all fears that are blocks to training and information assimilation are not impossible and can be overcome.

Anxiety

Anxiety can be an incredibly hampering factor in learning and in the application of force. Due to this factor's role in many elements of our modern lives, it has received considerable attention. Anxiety, or the emotional state where we feel uneasy, apprehensive, or fearful about an unpredictable or threatening situation is a common sensation and to some degree we all experience it when confronted with something new or a little pressure to perform at our best. However, the anxiety that we as use of force instructors need to concern ourselves with is that which is felt by an individual when faced with the need to act decisively under very arduous conditions, like a knock down, drag out fist fight or caught flat-footed gun fight.

Anxiety can be a severe impairment to performance throughout all forms of cognitive functions, including attention, memory and problem solving (Fiske, 1982). There is a documented interaction with the difficulty of a task and anxiety, as anxiety can result in poorer performance in complex skills, but may actually improve an individual's performance when undertaking simple tasks, remember the inverted-U? This influence on performance makes anxiety a highly relevant issue when developing realistic training. Our goal as instructors and leaders is not to do away with anxiety, but rather keep it at a manageable level that can actually enhance performance. Training needs to be developed that minimizes stress and actually prepares an individual to handle the situation. If training is successful, the student when confronted with an actual situation should have a sense of confidence and feeling of "having been here before". If instructors increase the use of positive or constructive feedback during training, then an individual's own anxiety about performing a task and fear of failure during the execution of task will also be greatly diminished.

Arousal

When an individual enters into a situation perceived as threatening, the body undergoes some very significant things, many we have already touched upon earlier in this text. One of the principal responses is the involuntary activation of the sympathetic nervous system (SNS) in relationship to an individual's sense of stress. It is important to recognize that stress means different things to different people, particularly in lethal threat situations and is a normal reactive response. Arousal is for all intense purposes the stress that we feel. It has been surmised through documented research that certain levels of stress are important to take us to the next level and let us perform at our very best (Inverted-U again) and there is a level of arousal that each individual needs at a given time to perform. If the level of arousal dropped below that level, the individual would need additional or outside arousal to continue to perform at optimum conditions (Berlyne, 1960). This makes a great argument for increasing the difficulty or challenge of training and not allowing an individual to fall in a rut or become bored with training.

Perception

Perception is generally the process by which we interpret and organize our experiences of the events in the world around us. We receive the information that we use through the senses of our body and mind (sight, smell, touch, hearing, and taste). Perception is directly related to all other higher-order cognitive functions like reasoning, problem solving, memory and motor skill behavior.

When discussing perception, it is important to realize that like many other elements of psychology there is a heated debate regarding the role of stimulus and experience in terms of which one influences our perception. One side concludes that stimulus is primarily responsible for what and how we perceive (Gibson, 1969) and the other side presumes that all perception is influenced by our own individual experiences, training and background. Many modern theories of social learning tend to support this position. In the realm of violence, receiving or applying, it is important to understand that the involuntary activation of the SNS and all its evils is based exclusively on the individual's perception of the situation. Further, an individual's perception is tremendously influenced by the type and level of training that the individual receives. The difference between seeing a situation as menacing, but manageable and intimidating and out of control rests in the individual's perception. We as trainers have a direct influence on altering an individual's perception. If we can create a positive, realistic and relevant training environment that is reinforced on a regular basis by the same type training, then we can have a significant impact on the individual's perception and his or her ability to resolve a situation appropriately and efficiently.

Dynamics of stress

We have all sat in classes where an individual, with a similar history as ours, "taught" us about the dynamic of a violent encounter. Have you ever asked someone who has never engaged in a violent situation, what it's like in one? Most can tell you that tunnel vision exists, but cannot tell you what it is. In the early 1980's, I was living in Northern Arizona and while running a small business, I trained in an outstanding local martial arts school. Well over a year into training the senior instructor became interested in full-contact bouts in Phoenix I and 2-3 others from the school began training for these fights. Following several weeks of very intense training, we felt that we were ready to compete. This was not my first time in martial arts competition. Several years earlier while living in Sacramento California, I competed in local and regional sparing tournaments and I lettered in wrestling in both Junior High School and High School. Having done very well at them, I felt I was competently prepared to fight full contact. I weighed all of about 140 lbs., but I was scheduled to fight in a higher weight class, which did not mean much to me at the time. We arrived at the convention center and I was blown completely out of my socks! First off, unlike point tournaments, there was a large capacity audience who was actually paying to get in and see the fights. The organizers took all the fighters into a separate room where we were weighed in and our hands were wrapped. Heavy weight gloves with sewn down thumbs were secured to our hands and a liberal coat of Vaseline was applied to our upper body, cheek bones and jaw line.

My turn came up, changed from the fifth fight to the first. When the announcer introduced me as having no wins, no loses, no ties and weighing in at a miniscule 140 lbs, I still felt fairly confident; but when he announced my opponent as having 16 wins, no loses, 14 knockouts and breaking the scale at a stout 165 lbs, I started to feel the onset of the SNS. My visual field got as tight as about a 10 inch circle at about 25 yards. My hearing went from crystal clear to a muffled garble and eventually to a dull ringing. Once in the ring, I looked around the room and saw all the people and the sensations of the SNS increased. The referee had us shake hands and commenced the bout. One minute and 30 seconds into the second round, I was knocked completely out. I woke up looking up at the lights on the ceiling and when I was told that the fight was over, the effects of the SNS began to diminish. I found out that although I had little memory of it, I had landed several punches that nearly knocked my opponent down, but I failed to follow up with additional techniques. I also found out that I had a loose tooth and a badly broken nose. The point is not that I was a bad fighter, actually in subsequent fights in the U.S. and Asia, I did very well, but rather that the SNS is an evil character that I was not prepared to deal with. However, I was able to learn from the experience and never experienced the SNS to the same degree again when confronted with a similar scenario. However, I have endured a similar exploitation by my SNS in the course of living through new experiences, including a couple of live fire skirmishes during my active duty military career and as a deputy sheriff. The point here is, as new experiences in life and in training are gained they alter or re-program us and change the overall effects of the SNS.

Mitigating Stress and Risk

There are many available ways to manage stress and risk in the work place. Unfortunately, wearing orange vests and hard hats cannot solve our ultimate workplace dangers. We are in a line of work, a lifestyle, that puts us face to face with a very bizarre hazard and as you can probably assume from the running theme of the text, the only way to mitigate this stress and manage this risk is through training. The first and most important step is to develop a state of mind that is generally void of emotion. Where perception, analysis and response become one fluid process.

The various psychological variables that are associated with a true winning mindset all center around developing an individual's confidence in his or her personal abilities and their reliance on the personal values and beliefs. Together, and working in unison, these elements can help build a mindset that allows an operator to focus on the mission without distraction and to respond appropriately without hesitation. We as instructors have a direct influence over all these factors, except for personal values and beliefs; these are very individual and personal issues and cannot generally be effected during the short duration of training. However, I can state with all confidence that values and beliefs play a significant role in cardiac and adrenal activity during high-risk situations. So long as instructors understand the relationship these elements have in a winning mindset, and in the over all mitigation of risk, they should be able to adequately express it to those in their trust.

Confidence, that "can do" attitude that helps build self-esteem and pride, is one of the most important aspects of mitigating risk. Through training, an awareness of an individual's capabilities is established. This awareness or confidence in one's abilities and the viability of a particular skill greatly minimizes the effects of the SNS and hesitation during moments of high stress. Remember, that the over all effects of the SNS are moderated by perception and perception is nothing more than a state of mind that is easily altered through training. I believe this is said best in "The difference in living and dieing is in the timing" (Lowery, 1985).

Another important element to this equation is the value of life. Now this may sound strange when the general premise of this text is training to apply force, but in reality the idea is to apply it in order to protect and preserve life. In our modern, ever changing society of growing complexity the value of human life remains a precious commodity. All the current faith systems, legal systems, and morals are structured around protecting life, just look at the incredibly heated arguments over assisted suicides and abortion. Due to the complexity of some individual's personal belief systems and the level of their commitment to the preservation of human life, there may be some individuals who have a great difficulty in having to take another's life, even to protect themselves. These individual's value and belief systems are so ingrained that the thought of killing is as foreign as suicide. This issue is interesting, because it is an accepted part of the profession that the potential to take another life is always present, and we voluntarily accept the responsibility to take a life, to protect another or ourselves (Smith, 1984).

To put in a nutshell, the most reliable way that we can help mitigate risk and stress is through proper training that literally guides the student through experiences that replicate real incidents.

Chapter Four

"If someone has a gun and is trying to kill you, it would be reasonable to shoot back with your own gun"

-The Dalai Lama

Use of Force Training

The typical concept of use of force training involves the act of repeating a skill numerous times until it becomes second nature or reflexive. In fact, many instructors believe that for an individual to perform a task in the field under combat stress, the individual must perform several thousand repetitions in training. This concept is generally motor skill development and it does help impart a program into the subconscious. However, imparting material is not the only problem. Anyone who takes on the responsibility of training others must not only apply good solid teaching skills, but must ensure that the material being imparted and assimilated by the student is relevant and applicable.

Traditional use of force training is actually

an exercise in motor skill development. This

type of training requires a relevant and

applicable relationship to actual situations.

(Courtesy of Perry Taylor)

In preceding chapters, I have discussed many physiological and

psychological limitations in use of force training, as well as the various

ways humans learn and apply learned skills. Among the multitude of studies

conducted in human performance, there exists three constant variables,

which bear a significant impact on use of force training. First, is the use and

development of a system of skills that are appropriate for their intended arena of use. Simply put this means that tasks and skills need to be designed to address specific stimuli. Second, the method of instruction must build a sense of confidence in the student's skills and abilities. Third, instructors must recognize that motivational methods and principles have a direct effect on the individuals training intensity and skills development. These three variables form the basis of training psychology (Siddle, 1995). Additionally, instructors must recognize their responsibility in helping to develop the student.

Current Thinking

A typical year in the life of any law enforcement officer is often said to consist of 364 days and 23 hours of boring complacency, with one hour being the wildest thrill rides known to man. The shear nature of the job lends itself to the development of routines and complacent activities. Through initial academy training and any continuing training received at the respective agency, the goal is to break the individual from failing into a routine, that is the goal but not always the end result.

Falling into a routine or becoming Complacent in response to a situation, like the often-fatal error of standing directly in front of a door as you open it, is all too common. One goal of training is to break these routines.

(Courtesy of SG Five)

The nature of the beast in modern law enforcement is periods of boredom and inactivity. Although mobile, the patrol car can be an attractive target for attack and difficult to fight from.

(Illustration courtesy of Mark Burnell)

Currently, the trends in use of force training don't differ much from 20-30 years ago. Typically, individuals face a single opponent, or in the case of firearms training a single target. The opponent usually acts or reacts to a set established stimuli and the individual is trained or conditioned to act upon that expected stimuli. The actions and reactions are generally choreographed movements, conducted in a controlled environment and are usually designed to be "win" situations. The training situation even starts from a predetermined position of advantage for the student. An excellent example of this is firearms training, especially low light training. The student stands on flat and stable ground with both weapon and light deployed and held in a "ready" position. Although this is a safe format to start in, it is not practical as their exists a gap between when the weapon is in the holster and the light is in the carrier and when they are drawn out and ready (I will expound on this further in an impending chapter).

Our Litigious Society

An odd state of affairs exists in our modern 21st century society; we have grown into a culture that is fascinated with the prospect of litigation and profiting from it. The daily television programming in the United States is saturated with countless court and judge programs that represent everything from petty small claims issues to full on ugly divorce themes. In fact we now have a stand-alone cable network that covers nothing but court matters. Among the incredible lawsuits over false complaints, landlord-tenant matters, and spilled cups of fast food coffee there is a tremendous amount of litigation revolving around the actual application of force or how that application is viewed or perceived. The act of training alone is often a heated topic, with an inordinate number of schools of thought, most which conflict or sits in total opposition of each other. In most organizations, the administrative entity is the official body that determines or allows what its members will and won't train on. Many times the administration permits its determination to be influenced by their perverse concern of public opinion and their fear of civil litigation, rather than from qualified expertise in the area concerned. Typically, administrative personnel, regardless of their professional roots, are usually concerned with budget issues, labor issues, and the creation of policy that will limit an organizations liability and financial involvement in civil issues that can occur as a result of the actions of their employees. The act of lawsuit prevention is most often conducted through inaction, rather than through appropriate training at an appropriate tempo. Even when training is used as a tool to reduce liability it is usually done to a point that is just enough to pacify some minimal standard and is always done as a reactionary response to an incident that has occurred. Unfortunately, most of the incidents that have brought about a change in training have cost someone their life. This method may limit, to some extent, an organizations chances of being sued, but places the individual officer, operator, or soldier in an extremely precarious position. Through the administrative lack of action, I like to call it the "brass two-step", the individual who is the end user, the soldier or police officer, is placed at a marked disadvantage. This disadvantage stems from a lack of internal support by the agency and a limit on resources that are available in times of crisis. This situation deteriorates considerably through the existence of many, though manageable and sometimes useful, physiological and psychological drawbacks in human performance. It is important to note that these drawbacks are inherent in all of us.

It is interesting to note the volumes of case law that have come about as a result of training related problems, particularly in the area of less than adequate use of force training. In the City of Margate (Popow v. City of Margate, 476 F Supp 1237, DCNJ 1977) a city police officer was in a foot pursuit of a suspected kidnap suspect. The officer was firing his duty handgun at the fleeing suspect as the two ran through a heavily populated residential area at night. Popow, an uninvolved resident of the neighborhood curious about the commotion outside of his home, stepped outside to see what was happening. Popow was subsequently struck by one of the officer's errand rounds and was killed. On a motion for summary judgment, the court found the following items established the City's culpability and established that the training programs were "grossly inadequate":

1. The officer involved initial firearms training took place 10 years earlier at the academy.

2. The only continuing training was shooting instruction, occurring approximately every 6 months.

3. The agency did not provide dim-light training.

4. The agency did not provide moving target training.

5. The agency did not provide training that addressed shooting in a populated residential area.

6. Problems existed in the City's policy and rules concerning firearms usage.

7. There were problems with inadequate supervision and discipline of police officers for misuse of force.

The end result was a lengthy civil trial and a subsequent settlement paid out to Popows survivors, as well as a literal re-invention of the training process for law enforcement

This need for changes in training and supervision was obvious, but not immediately acknowledged by all agencies. Nearly ten years after Popow v. City of Margate, in the City of Canton Ohio, another situation occurred (City of Canton Ohio v. Harris, 57 U.S.L.W. 4270, 1989). A female suspect was arrested and during the arrest she exhibited an extreme level of incoherence, marked by an inability to stand up or walk. The arresting officers summoned no medical assistance for the suspect and following her release, she was diagnosed with a number of emotional problems that required immediate hospitalization and treatment. The suspect eventually sued the city, claiming that her right to due process had been violated since she was not afforded medical attention. Evidence in the case showed that city regulations gave shift commanders sole discretion to determine if an arrested individual required medical attention, and suggested that commanders were not properly trained to make these determinations. Through this case the courts created a new term and standard when dealing with cases alleging problems with the training received by individuals and the programs administered to the entire organization. This term is known as "deliberate indifference". Although this case brought this term to notice, it was not fully established in this case. The courts held that the city was not liable under "deliberate indifference" for isolated individual acts perpetrated by lone officers. However, if agencies created or adopted training programs that were not relevant to usual and recurring situations encountered by their officers or the agencies were obviously indifferent to particular individual situations they could be held liable under "deliberate indifference". It is important to note that many agencies today still refuse to adequately address proper training issues. Solely basing this disinclination on the lack of budget money, public opinion about use of force training and that the incidents that have fallen on other agencies have not yet effected their agency.

These two blocks of case law are by no means intended to represent the cumulative amount of case law that relates to use of force training and litigation. Each day there is something new that is being brought before a court somewhere and, win or lose, these suites all set forth precedents for the next time someone chooses to sue. It is undoubtedly vital for instructors, regardless of your agency or discipline, to be functionally aware of the various case law and court standings that exist concerning training or the lack of training. A knowledge of these decisions, including their exact language, will help an instructor craft a program that will fill the existing voids, fit agency policy and promote appropriate response on the part of the student. However, it is not important for the student to have such a knowledge. I am quite sure that this will raise some eyebrows and may even launch some derogatory responses in my direction, but this is one, of many topics that I feel particularly strong about and one that I can relate some personal experience to. Several years ago, a portion of class that I was involved in called for a block of instruction on case law. The instructor stepped up to the podium and spent 3.5 hours covering a profusion of case law. The class, made up of primarily new officers, sat through over three hours of who got sued for what, how much it cost and how detrimental it could be to a career and a family to make a mistake and be sued. Three days later, while working graveyard shift with one of the young officers from the class, we encountered a combative subject while responding to an in-progress domestic violence call. The suspect, armed with a large kitchen knife and a softball bat was intent on causing serious injury on both my young partner and I. I watched as my partner hesitated and almost internally fought to respond, even after the suspect had struck him several times with the bat. Once the situation was over and the suspect was secured I asked my partner what was going on and he said that all he could hear and see was the class about case law. The potential of a lawsuit made such an impact in his mind that it actually interfered with his response to a threatening situation. The officer had been trained initially and subsequently to observe and respond appropriately, but found interference in that program when stories of how others lost money, careers, and families due to civil litigation clouded his thought process. This occurred even under the potential of getting injured or losing his own life. It is for these reasons that I strongly advocate teaching and conditioning students to "do the right thing at the right time" based off of appropriate and relevant training and not fill their heads with yarns of liability horror stories. As a trainer, a solider, a police officer and as a supervisor the last thing I want in the mind of someone in a threatening situation is something will cause apprehension. The shear notion of advising the working troops in any agency of the legal standards that are on the books in this country and the language that they are written in is ludicrous. By its nature, case law does not address the details of an incident that led to its finality. Case law is just case law. It addresses constitutional legal standards that are based from judicial interpretations of the U.S. Constitution. Case law is nothing more than a representation of how the courts view our actions and the actions of anyone else involved in the situation. This is not to say that I feel that case law has no place in training, quite the contrary. Case law represents an important component of training that is essential for the composition of legitimate and effective training programs. If properly used it gives instructors and policy makers and insight into how to do things better and train their students better.

The standard case law lecture session can be a valuable experience for administrators, supervisors and trainers, but can have disastrous effects when it is given to a work force of an agency as a supplement for actual use of force training.

(Courtesy of SG Five)

<u>Administrative Knee Jerk Reactions</u>

Several years ago, during the "wars" that were waged between U.S. manufacturing industries and those of Japan a number of significant details were revealed about our general philosophy concerning mistakes in the work place. The most noteworthy of these details was how both cultures viewed a problem and how they went about solving it. In the United States, generally, we discover a problem or a mistake in the workplace and before we solve the problem we try and learn who is at fault or whom we can blame. Once the guilty party is identified and chastised we try to fix the problem. In Japan they view the situation somewhat differently. Once discovered, the Japanese do their best to fix the problem and resume productivity first. Once the situation has been brought back to normal an attempt to determine how the problem happened and how to avoid it in the future is set forth. As the whole process draws to a conclusion they determine who made the mistake and ensure that the particular employee is counseled and re-trained accordingly. The philosophy of attaching blame first is very evident in the world of our military and police agencies; this is the typical administrative knee jerk reaction to a problem. Instead of looking at the big picture and trying to solve the problem, we have a habit of pointing fingers and creating policy or firing employees.

Several years ago, in an agency I worked for in Arizona, we were allowed to carry any type of sidearm we wanted, so long as it passed an armorers inspection and we could qualify with it. A senior officer in one of our more metropolitan districts had opted to carry a personally owned single action 1911. One particular evening the officer was investigating a reported burglary in-progress call at a local car dealership. The officer arrived on scene, drew his pistol and his flashlight and commenced to clearing the lot for any possible suspects. All of a sudden a large naked man jumped up from behind a vehicle and began to scream and jerk incoherently. This action startled the officer and caused him to accidentally discharge a 230-grain round into the rear quarter panel of a new car. Long story short, the administrative knee jerk to this situation was to immediately suspend the authorization for agency personnel to carry single action pistols. This decision was made without due regard to the totality of the situation, it was simply a quick fix and an ineffective effort to prevent being sued. A subsequent investigation into the situation revealed that the officer had been searching with his finger on the trigger and when he was startled he squeezed and discharged the weapon, a simple and common mistake. Countless pieces of information were provided to the administration that showed that just as many accidents have occurred with other systems of handguns as with 1911's and that the common denominator in the majority of these accidents was a finger on the trigger. Ultimately, the administration allowed the carry of the "malevolent" 1911, but only by "senior" officers who had attended a 40-hour block of remedial firearms instruction. The conditions imposed on the agency did not address the actual problem. As a senior officer had been the one at fault in the first place, suggesting that time on the job had nothing to do with the accident or in preventing it. Further, a 40-hour class on remedial firearms training was a standard class that did not address the safety precautions of using a single action pistol. All in all nothing was really done to address an officers problem in weapons handling skills and search procedures, but the administration reacted in a fashion that was very typical and one that we as instructors need to be aware of.

Beware, The Self-Proclaimed Expert

Training, regardless of whom you work for, is without a doubt the most important function we can engage in when not working in our primary role. Training is the vehicle that prepares us, our students, our supervisors, and yes, even our administrators to perform their jobs and to win in all situations. Training provides us with an authoritative foundation for both the individual and the organization. Any agency, despite size, mission, or authority must be trained and maintain a reasonable state of readiness, prepared to protect lives, deter hostile actions, control disturbances and to terminate special situations in difficult and undesirable ways.

This need for training and preparation demands that training be set to standards and that all personnel involved in training understand, attain, sustain and enforce those standards. To fully prepare and develop an organization, tough, realistic training designed to challenge the student must occur. To supplement, sometimes compensate, their in-house training programs a great number of organizations rely on the numerous private trainers and training organizations that exist today. Most of these trainers come from a competent pool of seasoned, military and law enforcement veterans, teaching from a foundation of practical experience and knowledge, unfortunately many do not. There are those "elite experts" who base their skills from the books they've read or the videos they've watched and have never been in nor witnessed a high risk situation, yet they "teach" systems that are supposed to equip a warrior with the tools to prevail. This situation is like a virgin teaching a class on sex. Investing time, money and lives in the hands of an ill-qualified "expert" can be a fatal mistake. Now is a good time to state that I am not bashing private trainers or training organizations, if I was I would be bashing myself. I have worked with a number of very fine trainers and organizations and most provide your moneys worth. It is important for an instructor to understand that there are folks among us who are not qualified and can actually cause damage; this is the primary focus of this portion.

All too often ill-qualified instructors waste training time padding their own ego and

getting students hurt.

(Courtesy of Perry Taylor)

The later part of the 20th century brought a veritable explosion of "expertise" to the world of tactics and use of force training. Never have so many specialists, real or otherwise, made so many claims concerning their expertise and knowledge in regards to the reality of high-risk situations. Many of these individuals have gone that extra mile to see that their opinion is well known and their pockets are well lined. One only needs to peruse the advertisements on the Internet or in the back of any popular magazine to get a classic example of this phenomenon. Usually in bold print with vivid photographs are ads claiming to teach techniques that are exclusive to one of many special operations organizations. Advertisements alleging to be the "founders" or "creators" of techniques and styles that will turn the meek into super warriors. These ads are usually accompanied by an impressive, but unbelievable, list of titles, credentials, and comments made by students who, for "National Security reasons", must remain anonymous. That's the same story that Bugs Bunny uses on Yosemite Sam.

While there are certainly many outstanding instructors and training organizations pursuing the development of newer and better programs with worthy intent and genuine desire for excellence, there are an alarming number of individuals making outlandish affirmations about their personal deadly efficiency and the effectiveness of the techniques they developed. It is typical for these so-called instructors to allege very close ties with specialized military and law enforcement organizations, often claiming to be the "official" instructor for the unit or organization. This behavior can only be explained as a deep psychological and social need to sing one's own praises and use it as a method of excluding one's self from others. Although this activity is not an absolute male endeavor, one only needs to listen to little boys on a playground posturing, trash talking and picking fights with each other to discover these issues are prevalent. In fact, many of today's popular television talk shows, reality based shows and au courant "gangsta" rap songs are centered on man's propensity to shoot his mouth off and ridicule the abilities of his contemporaries.

Many of these self-proclaimed tactical masters openly promote extremely violent training programs and equally violent responses to situations. In their own morality, they often encourage the swift and immediate use of deadly force to any and all conflicts, without regard to intent, the factors of the situation or the location of innocent bystanders. The world of reality today and the world of training, especially for law enforcement, is one of profound scrutiny. The application of force has a tendency of upsetting onlookers on scene as well as later on the evening news. Having no real frame of reference for the situation, the public is generally influenced by their personal emotions when evaluating these incidents. A trainer or training organization that fails to realize these issues is hopelessly ill prepared to conduct training of any real value. Additionally, these "experts" cause a tremendous cycle of damage when, because of their larger-than-life credentials and synthetic backgrounds, self-proclaimed experts are viewed as authority figures, especially by the more naïve of their followers. As such, their position of influence and strength lends a staggering sense of validity to their actions. Further, the self-proclaimed expert often displays a behavior pattern similar to former Vice President Gore, in that they want everyone to believe that they and they alone discovered or invented techniques or systems. Like Al Gore and the Internet, self-proclaimed experts lay claim to many popular mechanisms of use of force training and are often heard whining, nagging and complaining about others who have stolen their "exclusive" material.

Professional trainers are both teachers and students, they must believe in their role as a teacher and be a lifetime student. An instructor is successful only to the extent that they enable their students to learn what they need to know, measuring their success in those results that come from the instruction and the student's performance. Professional instructors must be highly competent in the subject matter that they are teaching, but be leery of ever referring to themselves as "experts". Well known trainer and author, Gabe Suarez, has often used the term "advanced student" in lieu of expert. Personally, I think this is a very befitting title that truly says it all. Myself, I am in a continuous learning cycle and I don't think there has been a class yet that I've been involved with where I haven't learned something from one of the students, if nothing else a little humility. Our role as a trainer should never stand for the all knowing, supreme, master of our subject, rather it should represent a level of expertise in our particular discipline and our quest for greater knowledge and understanding is never complete.

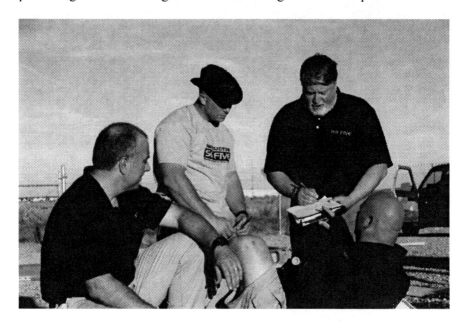

Its very common for instructors to learn from students during the course of a training session. While we have a level of expertise, the title of instructor really means "advanced student", not expert.

(Courtesy of SG Five)

Use of force and tactics training is a tremendously complex process that is not always understood by the participants or by the instructors. Relying on the purported experiences of self-anointed prodigies or the advice of others who are equally misinformed can be tragic with far reaching complications. Far to often we find ourselves relying on our perception of a trainer to assess the "truth" in his or her statements, often placing our trust in people who exude high degrees of charisma and confidence. Jim Jones was a very charismatic minister who leaked confidence from every pore. Jones convinced hundreds of men, women and children to follow him to Guyana and to commit one of the largest mass suicides in history. Further, we gullible humans have a tendency to believe that knowledge is possessed in proportion to the diplomas we hang on walls, the size of a resume or the number of books someone writes. Keep in mind, the scarecrow in the Wizard of Oz wanted a brain, but got a diploma instead. In simple terms, you cannot determine an "experts" worth from what he is or claims to be. These statements are not intended as encouragement to hurl unsubstantiated accusations against everyone in the training business because there exists a great number of knowledgeable, high-energy, high-productivity folks who truly strive for excellence and do a stellar job producing an outstanding end product.

"Excellence is an art won by training and habituation. We do not act rightly because we have virtue or excellence, but rather have those because we have acted rightly. We are what we repeatedly do, excellence then is not an act, but a habit."

-Aristotle

Chapter Five

"Knowing is not enough; we must apply. Willing is not enough we must do"

-Goethe

Reality

Reality, What is reality? Is it our day-to-day world of life, the things we see and the things we can touch or is it only our own individual perception of that world? Pretty deep, huh? Reality, although a word viewed with different definitions, is truly a conjectural quandary and is actually one of the basic questions of philosophy. Just like many academic disciplines, including education, reality is subject to the opinions and theories of many entities. There are those who believe that reality is solely a product of the human mind and there are those who believe that reality exists apart from the human mind, or within nature itself. The Buddhists, for example, see reality as what they refer to as "The four noble truths":

1. Life is suffering, or the recognition that life is a painful journey from birth until death. Even in death, Buddhists, believe that suffering does not stop, because of the belief that life is a repeating cycle and that death simply leads to rebirth.

2. All suffering is caused by ignorance. Primarily ignorance of the nature of reality and the desire, attachment, and selfishness that comes from that ignorance.

3. All suffering can be ended by defeating ignorance.

4. The path to ending suffering is the Noble Eightfold Path, which consists of right views, right intentions, right speech, right actions, right livelihood, right effort, right mindedness, and right contemplation.

The Greek philosopher Plato, responsible for great influence over western philosophy, saw reality as two distinct levels of awareness: opinion and knowledge. Affirmations about the viewed world, including valid observations and scientific proposals, were just opinions and regardless of their foundation, none were considered actual knowledge. Plato considered a higher level of awareness to be actual knowledge because it required reason and reason required intellectual insight that was undeniable. Further, Plato reasoned that this insight gave a distinct and accurate view of the forms and substances that constituted the real world or reality.

Plato's concept was best described in a story that he told and like any other profession the use of stories and analogies helps to describe the material. This parable involves a group of men bound in a dark cave. These men could not see each other, but could see the shadows of figures and statues on the walls of the cave thanks to the flickering light of a small fire. Eventually, one of the men escapes the cave and finds away into the world and for the first time actually sees things the way they are. The man returns to the cave and tells the others that all they have ever seen are nothing more than shapes and shadows. The man tells the others that another world awaits them if they are willing to struggle free and escape. Plato symbolized the physical world of appearances with the image of the dark shadowy cave and the escape from the cave stood for the transition to reality and to knowledge.

Now, understandably this is a bit deep and not entirely applicable to use of force or martial training. However, as previously noted in this book, the mind cannot always tell the difference between reality and fantasy, thus leaving reality up to one's own perception of his environment. This mind-body connection and our ability to play upon reality gives us a tremendous amount of latitude when creating a positive training environment and create controlled environments where we can actually influence an individual's perception and condition them for appropriate responses, more on this later.

Actual Reality

One of the primary goals, or at least it should be, of an instructor is to create a realistic environment that, although controlled, replicates actual conditions that the student will face in their given profession or lifestyle. A replicated environment, if planned correctly, can do a great job in preparing the student for the reality of the streets because actual reality is employed in the mock environment. For example, in the land based forces of the United States military (U.S. Army and U.S. Marine Corp) they depend heavily on an individual's ability to navigate by using a map and a compass. Even with the growing use of military grade GPS systems (PLUGGER) the old fashioned methods of land navigation are still a staple skill of basic soldiering and must be mastered before graduating basic training or boot camp. Following classroom instruction the students are taken to a physical land navigation course where, either alone or in two man teams, they must locate coordinates on a map and physically travel to them with 100% accuracy. The land navigation course is a simple but highly effective mock environment. The students are placed in a location, usually covering several square miles, and must correctly navigate from point to point on foot or mounted in a vehicle. The course accurately replicates basic real world conditions, as the students are usually in load bearing gear, with weapons, helmets, and rations. The only thing missing is aggressive hostile forces (that comes later if your lucky enough to go to SERE training).

Wes Doss

Replication of real life situations, like shooting from unconventional positions or

with a protective mask, is crucial to the conditioning of an individual to respond to

spontaneous situations.

(Courtesy of the author)

106

The land navigation course is a simple representation of a realistic training environment. However, it is important to note that no replicated environment can actually produce the same stress and conditions of the real world. The training environment, no matter how realistic, still misses the spontaneous and unexpected circumstances of actual conflict. It is quite possible to approach these conditions but because it is training and because we don't want to hospitalize or kill our students, we have to pull our punches a little. We can introduce small doses of spontaneous behavior into our training milieu, but anything that lasts uncontrollably for a length of time is likely to end up in serious injury or damage to property. This is not to say, in any way shape or form, that I do not advocate realism. In fact, realism is the real bases of this book and an element of training that needs to occur more frequently. When we step up the tempo of training and we increase the level of realism, we multiply our safety problems considerably. The instructor must assess the student and the situation constantly and determine if the level of training can be safely stepped up or not. Additionally, the instructor or instructing staff must ensure that appropriate safety equipment is present at all times. This equipment will vary depending on the type of training being conducted, but generally should consist, as a minimum, of the following:

- First aid kit (with general life support considerations)

- Ice or ice packs

- Water

- Fire extinguisher

- Communications (at least two forms)

- Transportation

As we increase realism in training we also increase the potential for injury. A

correct balance of safety and reality is vital.

(Courtesy of SG Five)

Far to often instructors try to start out running before they are sure the

student knows how to walk. Even active members of highly conditioned

units and organizations are going to have individual differences in abilities and skills. It is important to remember that each of us is different. Go to the mall this weekend and just look around if you don't believe me. We are definitely not a one size fits all world!

Forced Reality

The concept of forced reality is where the rubber starts to meet the road. Imagine the benign land navigation course described in the previous section, only this time imagine gun fire, vehicle traffic, pyrotechnics and hostile opposing forces. When we introduce unprompted components of realism to our training environment we are applying the principle of forced reality. This particular action increases our safety consideration ten fold because we have now introduced elements to the environment that can cause additional injury and up the potential for fatalities. However, the value of adding spontaneous events to our setting in invaluable, even if its nothing more than background noise played over a loud speaker.

The value here is in the creation of a training area that most accurately replicates the things that a student will find in their world. When applying forced reality we want to try and accurately reproduce the sounds, sights, smells, tastes and verbiage that will be encountered in the real world, yes verbiage! Much to my ongoing amazement, I have been involved in numerous heated debates concerning the use of language, foul or otherwise, in both the training and operational environment. Apparently, there is a vast portion of our adult population that has never heard another human utter a coarse word, either out of anger, upbringing or in jest. It's hard to visualize this because the majority of the folks I deal with are big tough guys that are actively involved in some form of violence-oriented occupation. I can only assume that this reaction stems from the deep influences that "political correctness" has on all elements of our society. Now I don't encourage the undisciplined use of foul language, especially in classroom or professional environments, but certainly if we are trying to replicate a hardened drug house full of hardcore gang bangers we don't expect to hear politically correct terms being uttered. The shear fact that people object to this language and find it disturbing makes it that much more important in a forced reality environment. When we introduce unexpected effects into the environment we are creating disruptions or distractions. These diversions are sticking points in our development that we need to condition students to go beyond and if offensive language makes you stop in your tracks, then it is definitely a disruption that must be overcome.

Role Playing

With the introduction or implementation of forced reality it is important to not overlook the one factor that can make or break the training, the human element. One of the most difficult components of law enforcement and military careers are having to interface with other human beings. In fact, most people find it easier to contemplate inflicting injury on others than they do having to speak to a stranger, much less having to interface with one in a hostile setting. By using role players in our training scenarios we increase the realism and problem solving components tremendously.

A credible and realistic role player can elevate a training session to very productive

heights.

(Courtesy of SG Five)

One of the most neglected facets of using role-playing is utilizing credible role players and drafting scenarios that allow for realistic actions and outcomes. Often members from the same team or organization are used as role players in training, while this is more or less effective economy of manpower it does not provide for a realistic situation. Too often, training scenarios pitting teammates against each other end up in win-win or lose-lose predictable formats and play out to unrealistic Hollywood type finales. There is generally too much familiarity with tactics and techniques among teammates to get any real benefit from the experience and can often end up in a day of playing "grab ass", getting little to nothing accomplished. It is also difficult to be serious when you get nose to nose with someone you work with on a daily basis. All it takes is one silly grin or a giggle and the entire situation is down the tubes. The Department of Defense realized quite sometime ago that to glean the maximum benefit from hard realistic role-playing exercises a separate element trained in unfamiliar or unconventional tactics was needed. Since this revelation a number of dedicated opposing forces units or OPFOR elements were created. These specialized units dress differently, train differently, execute tactics differently and use equipment that is distinctly different than that of the rest of the military. When an OPFOR unit is used in an exercise, there is no predetermined winner of the engagement, even though most OPFOR units practice on the same battlefields where they deploy giving them a home court advantage over their opposition. This advantage is as well part of the realism, since our armed forces seldom deploy to hostile territories within our own borders and our opposition is typically on their home turf, this is true for the military as well as police. The specialized OPFOR units are obviously not cheap and for understandable reasons most of us and our employers cannot afford the time, money, or manpower to create these elements. However, there exist a great number of outstanding sources for credible role players, including; high school, college, or local drama groups and members of other agencies. Anyone or any organization that is generally outside your own, and is willing, can help fill the void of role players. I have personally been involved in a number of successful scenario based training exercises that were designed around the concept of a school or work place violence incident. In these exercises actual students and staff were used as role players. The fact that they were completely unfamiliar with the tactics and techniques of the participating agency made the exercise very spontaneous and productive. In fact, because actual staff was used, a number of undiscovered deficiencies in the facilities emergency plans were detected and were ultimately changed as a result of the exercise. However, it is important to remember that if you opt to use civilians and especially under-aged civilians in your scenarios, you need to ensure that proper liability and responsibility waivers are taken care of. After all we are training to manage and control hazardous situations and occasionally things can get a little dicey, even in training. Even though I firmly believe in hardcore and often extreme training I also believe, just as firmly, in minimizing injuries, especially stupid unnecessary injuries.

Simulation Systems

This tome is not an advocacy or endorsement for any particular product and in this section I will limit my references to any specific system and I will reserve my own personal feelings for another time. In discussing simulations systems it is important to note that regardless of your beliefs about them they do fill a distinct void in the training spectrum and can be an extremely constructive tool in your training program. There are numerous simulations systems and software that exist and more are being produced everyday. In fact, the information technology field, specifically the simulations and gaming field, is growing tremendously. The idea behind the simulations system concept is to accurately produce a benign environment where actual techniques, tactics and equipment can be employed. The aerospace industry and the U.S. Military have been successfully using simulators, of one form or another, for a number of years. Computer generated images of battle tanks and aircraft have been a part of our world for many years. Within the confines of this book, and for all future reference when referring to simulations systems I will be discussing systems that are applicable to individual or team operators, who generally contact their opponent at relatively close distances and are in use by a great number of law enforcement agencies and military ground forces (FATS, AIS, I-Sim, etc.).

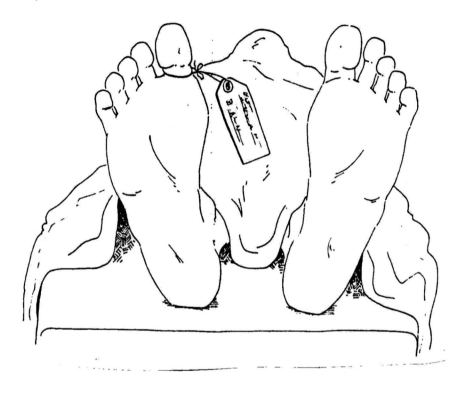

The use of solid realistic training can help prevent unwanted tragedies.

(Illustration courtesy of Mark Burnell)

These systems are generally some configuration of computer, media player (laser disc, DVD, video tape, or software) and a large interactive screen that work in conjunction with a specially configured weapon that the system is able to communicate with. Some simulations systems allow for the use of actual duty weapons and live ammunition and others are simply a dedicated weapon that is dry-fired. Regardless of the system or the configuration, simulations systems allow individuals and teams to interface with a "live" target and escalate and deescalate through the full force continuum before making the decision to apply deadly force. They allow instructors to watch a student closely and help coach the student through the scenario; some systems even have a target-shooting mode that can help diagnose problem shooters. The use of simulations systems are relatively safe and cost effective, but their use, like all training tools, must be kept in perspective. Far more often than I really want to acknowledge, folks responsible for training make the critical mistake of considering simulation systems the supreme end all-be all of training. Any simulation system is just that, a simulation. It is a device or tool that is intended to augment other forms of training. Alone, a simulation system cannot solve all your training needs. In fact, it can't even come close but there are those out there who believe it can. To avoid confusion and stepping on more toes than I already have, please understand that I resolutely believe that the various simulation systems have carved out a niche in the world of training and can be a tremendous addition to any training program; but like any other element, they have to be used appropriately to glean the maximum benefit.

What Simulators are and what they are not.

The average stage for simulation devices is a cool, dark, air-conditioned room or trailer that, except for the dialog on the computer program, is void of outside noise or distraction. There is usually a briefing by the instructor and another prompt by the program itself before any "action" occurs. The student stands in a more or less static position, occasionally given the option of taking cover behind items that have been placed between him and the screen but generally restricted to a small area. Depending on the weapon used and its configuration, the scenarios can start with the weapon in the holster or with the weapon deployed; but either way, the student is told in advance. If the scenario involves a night setting, the student can hold a flashlight but cannot turn it on because the light will diminish the system's ability to read information from the screen or cause false readings. Some systems have addressed other components of the use of force continuum and allow for the integration of simulated chemical aerosols and batons but don't have the over spray, weapon retention, or control problems of the real world, thus thinning the realism of the program and reducing the student's anxiety or stress level. There are some systems that increase the stress and anxiety levels by giving the instructor the capability of remotely firing rubber balls or marking projectiles at the student in conjunction with the scenario being played on the screen. This arrangement makes the scenario seem more interactive since it will actually shoot back. However, the scenario still starts and stops at predetermined and predictable points with a general duration of no longer than one minute. Additionally, the scenario being played out and the movement restrictions in the training area work against the hard drilled concepts of distance and cover. Although the manufacturers of these systems often maintain an extensive library of captured scenarios, most agencies because of the cost of the media material, only have one or two sets of scenarios. After a few training sessions most personnel learn the scenarios and are not stressed or surprised by what plays out on the screen, bringing us back to a training day of playing grab ass. I think that's enough drawbacks, each mentioned really only matters in organizations that rely too heavily on the simulator and have stopped considering it a tool.

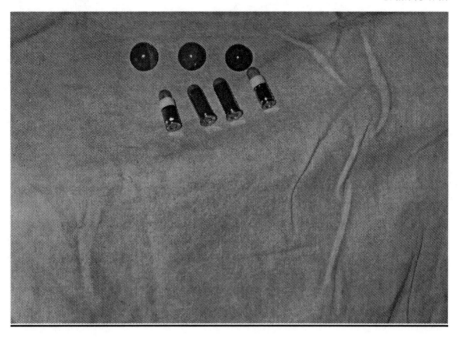

A comparison of Simunitions 9mm & .38 Spl. cartridges and .68 caliber paintballs

(Courtesy of the author)

The real usefulness of a simulator comes from running a supervised scenario with a running dialog of feedback where the instructor observes and constructively critics the student's decision-making capabilities. Most scenarios that are recorded are a careful examination of a very complex scene. As the scenario runs, it is very easy for the student to get swept up into the situation, waiting for the chance to shoot and may overlook some very significant and subtle cues. Sit back and watch a scenario taped by any of the major simulator manufacturers and watch the people in the background or the reactions of the simulated bystanders. Most of these

119

scenarios have people wandering in and out of a danger zone randomly, just like the real world. Each of these extras in the scenario is a cue to increase or decrease the level of force or select a different tactic. Following a scenario, the instructor and the student can discuss and justify actions, discuss alternative courses of action. In this format, simulators are invaluable. Just like the pistol on your belt or the hammer in your garage, simulators are tools. They have a unique and functional role in our training and can be a tremendous asset in conditioning individuals to make reliable decisions. If you ever encounter someone who feels that simulators are more than tools consider this: the hammer is a great tool, but it takes more than a hammer to build a house.

Force on Force

If forced reality is where the rubber meets the road, then force on force training is where the road meets the autobahn. Another concept that has found years of validation with our armed forces and many law enforcement agencies is the concept of force on force training. There exists no greater a tool to the instructor involved in martial type than that of force on force training, because it is the surest and most reliable method for increasing stress and applying actual techniques in a very realistic environment. In fact, because of the variety of products available for force on force training it can be conducted literally anywhere and the injury rate can be very controlled. Generally, force on force training involves the use of actual weapons and either marking projectiles (Simunitions, Airmunitions, Code eagle, paintballs, etc.) or low-level laser scoring devices (MILES, laser tag, etc.).

While safety is always a primary consideration in training it is also important to keep the stress and complications of conflict in perspective.

(Courtesy of SG Five)

121

Marking projectiles are what I find to be the most valuable in force on force training. With the exception of paintballs, marking projectiles are used in actual, but mildly modified duty weapons. This use of actual duty weapons skyrockets your realism, allowing for all conventional methods of carry and even account for reload and stoppage drills. Additionally, this mode allows this type of training to adapt quite successfully to the needs of a variety of students, bodyguards, police, military, and civilian. Paintballs, specifically the 68-caliber grade, are also extremely useful in this type of training. Although the weapons are not as realistic as duty weapons, paintball weapons do offer some serious advantages over the array of duty type marking projectiles that are available. Beyond the lower cost, the 68-caliber paintball weapons offer a marked level of down range trajectory and velocity well over and above the other items on the market. This increase in accuracy and velocity means that you can expand your training area to greater ranges, but safely collapse it to close quarters when called for, adding realistic flexibility to your training. There are brands and configurations of 68-caliber paintball weapons that are capable of reliable five to ten inch groups at out to 50 yards, still maintaining enough velocity to break the projectile on impact. Regardless of the type of marking technology you choose to use, they all create the undeniable trademark "splat" of paint, dye, or colored soap when you are hit. This obvious mark greatly reduces the potential of arguments over who was hit and who wasn't.

Beyond the visual recognition of being hit by a ball of paint and wearing the spoils of war all day, I am particularly fond of the pain value that force on force creates. Regardless of what type of projectile or device you are using, they are all projectiles that exit their respective weapons with enough force to cycle the actions and travel their intended distance. Therefore, all marking projectiles have a tendency to hurt upon impact with living tissue. The pain recognition characteristic of marking weapons may or may not have been intentional on the part of the developers, but it is undoubtedly a valuable resource. The use of pain as a motivator is an outstanding tool; it vividly shows shortcomings in tactics and exploits mistakes that were made. It gives the student a mental reference to what went wrong and can greatly reduce the likelihood of show offs or cowboys. The problem with the issue of pain is many people tend to armor themselves heavily from head to toe in an attempt to avoid the pain, but without the pain and more specifically the fear of the pain we greatly diminish our training value. Although, I am not fond of being hurt, I am a firm believer in the concept of using pain as a teaching tool and would encourage trainers to limit the amount of armor that students are allowed to wear during these training sessions. Further, additional armor or padding is unusually restrictive and unrealistic distracting further from the goal of quality training. Please don't misunderstand, I am against over doing the padding, but am in full concurrence of utilizing reasonable safety equipment to prevent unnecessary injury and to protect the eyes, throat and groin. At a minimum, the following should always be employed when engaging in marking weapon force on force training:

- Pants (not shorts)
- Eye protection (wrap around)
- A cup
- Paintball face mask and/or throat padding

While taking into consideration the various safety considerations of force on force training, we have to keep the training concept in perspective. The whole idea behind force on force training is to correctly replicate the various stressors and complications of actual conflicts and engagements. The more elements of an actual attack that are present during training, the more assured we can be of future success (Good, 1999).

Putting it Together

To have a truly successful training program you cannot rely on classroom or static application of techniques. At some point, in order to cultivate a winning behavior, a transition must occur linking the techniques and procedures learned to actual situations. A training environment is a much healthier environment to produce this transition in than an actual engagement. When realism is added to a training program, and then eventually force on force training, the training shifts to hard and demanding guidance, but then again the real world is, or more, hard and demanding.

Any realistic training situation requires a high level of supervision, detailed planning, and a relative level of expertise on the part of the training staff. The potential of accidents and injuries has to be mollified through control of the environment and of the conduct of the personnel involved in the training. A distinct gray area exists in the area of control, on one side you want to maintain as rigid a control as you can and on the other side you don't want to diminish the training value by causing unnecessary interference or distractions. When a scenario has obvious referees and support personnel running around or when training halts after every action to do an after actions review (AAR), it takes the realism down a notch or two and it stales the tempo. Support personnel or evaluators should be neither seen nor heard, until it is time or they are needed. Further, AAR's don't need to be conducted after every single action, allow the scenario to run until completion and then conduct a review. I have been involved in both large and small-scale exercises that did AAR's immediately following every action and in ones that ran on 12 or 24-hour clocks and did the AAR's at the end of the training period. The exercises that ran through to completion maintained a steady and realistic tempo, as opposed to having to stop every so often to talk about what we just did.

When running any type of training, especially realistic training, planning is paramount. A detailed and thought out scenario lets role players and participants know what is expected of them and what actions are appropriate for the incident at hand. For example, we want participants to handle a hostage rescue exercise differently than a protective services exercise. The scenario or script for the exercise should be explained in adequate detail and should include a description of events that led up to the scenario itself and the status of the immediate situation. In addition to the descriptors of the situation the planning should also include some semblance of rules of engagement (ROE). The ROE's should be simple and clear, yet give the participants a lucid understanding of what they are doing and what is appropriate.

During the actual implementation of an exercise, the trainer or training staff must maintain absolute control of the environment and its inhabitants. It is unbelievably easy for participants, particularly undisciplined participants, to get carried away and push beyond the limits of the exercise. This tendency rears its ugly head for the most part when tempers get raised or frustration over questionable performance comes up. This is not to say that we don't want to push the envelope in training. Remember that this type of training is a skill developer and, just like weight training, we have to push beyond our comfort zone occasionally in order to progress or to excel. It is vital for the training staff to expect individuals to "push it" once in a while and with that expectation in mind, give a little latitude to the students but keep control of the situation and get ready to pull the plug when the circumstances become out of control. There is a tendency among role players, especially untrained or overly familiar role players, to go off track and get carried away. Whatever source you use for your role players, you must ensure that they are well rehearsed and stay within the boundaries set forth in the scenario. Conversely, you have to watch participants reactions and consider the safety of the role players, as undisciplined participants will also have a tendency to get carried away and may cause undesirable injuries to members of your staff.

The use of realism in training for conflict is absolutely vital; the complications that can result from ill-prepared operators being thrust into the real world are staggering and senseless. No other element available, except the mind itself, will help you prepare others to win better than realistic training. The shift from all other forms of reality based training to full on force on force training can bring surprising results. Force on force can better prepare an individual for violent encounters and it can also expose current deficiencies in training and preparation. It is important that students understand this and don't fear it. Identifying problems and making corrections is a key element of training. However, a considerable amount of preparation must happen before the implementation of this type training. Additionally, all involved must be prepared to manage the training environment; otherwise all the time devoted is wasted.

Chapter Six

"Before you become too entranced with gorgeous gadgets and mesmerizing video displays, let me remind you that information is not knowledge, knowledge is not wisdom, and wisdom is not foresight. Each grows out of the other, and we need them all"

-Arthur C. Clarke

Environmental Training

Wherever you are, that is your environment. We all exist in and move throughout an environment and depending on where you are or what you are doing, the environment will differ. However, current trends in training would have us believe that all environments have level unobstructed ground, quality lighting, and outstanding ventilation. The reality of our various environments, including our own homes, is decidedly different. Our lives are full of clutter, blind spots, dark areas, stale air, and angles. As our world grows more complicated and aesthetically oriented our environments become less manageable from a tactical standpoint. Additionally, in sharp contrast to many popular training methodologies the various tactically ill-efficient components of our environments cannot always be avoided.

Environmental training, when discussing martial type training, is essentially an exercise in the use of reality and an attempt to heighten an individual's awareness of his or her surroundings. Current or standard methods of administering use of force training do very little to address this issue. In fact, there exists both policy and practice in many organizations that discourages actions that are conducive to awareness training. For example, an officer called to a disturbance in a large chain home improvement store. Upon arrival, the officer encounters an agitated man and during the encounter a struggle ensues. The officer draws his baton and attempts to apply the infamous, and often futile, common peronial baton strike on the man but looses his baton. The officer desperately tries to recover his weapon as the man grabs a hammer from a nearby shelf and pummels the officer until he is unconscious. Throughout the struggle the officer, although surrounded by potential weapons, tries desperately to recover his baton, even while being beaten. The officer's training to this point had not addressed the use of other items as weapons or the benefits of being aware of the suspect's access to weapons. It is quite possible that policy from the officer's agency had strictly forbidden the use, under any circumstances, of anything not issued as a weapon, as this is a frequent situation in many of the more "politically correct" agencies.

Learning and developing a skill, but never learning when or why to employ it in a situational status means that the student will most likely not be able to recall or use the skill when under duress. The information may not have been reliably assimilated and it may not transfer over to a response in a real situation. If we don't reliably transfer information to our students, then what purpose have we served?

As previously discussed in this book, learning occurs on many levels, but for our purpose in the field of force application we are going to concern ourselves with two; motor skill level and decision making level. The human brain learns specialized and synchronized physical actions and stores them as motor programs. When appropriate and facing specific stimuli, assuming that proper training was conducted, the motor programs prompt various nerves and muscle fibers to perform in response to the stimuli. The brain exploits a different apparatus to learn the components for making decisions, including decisions about motor skill responses. Essentially, the brain learns, stores, and recalls physical techniques separately from the decision making process and the understanding of the situations in which the physical techniques will be used. This is an example of why it is paramount for our students to manipulate equipment and function weapons from a variety of positions, as static set shooting positions will not prepare them to operate in an actual engagement. The operator must use strategy and supplementary techniques to make things possible. Additionally, the operator must learn to assess the situation, recognize available and viable tactics, determine temporary goals and understand ultimate goals in order for their brain to make connection and choose the right motor skill, assuming that they have been properly conditioned. It is important to note that some courses of actions are like a fail-safe self-destruct mechanism; once they are started they can't be stopped. Because of the time it takes to acknowledge and respond to a specific stimulus and then decide to abort it, is considerably longer than the time it takes to complete the chosen course of action itself. A baton in motion and centimeters away from impact and pulling a trigger are prime examples of this. Once these actions are initiated and in motion, they cannot be stopped. This is a fair example of Guthrie's contiguity theory and the relationship between stimulus and response.

High risk encounters often require an individual to apply more than one motor skill

at a time.

(Courtesy of Perry Taylor)

Unlike other occupations, those that deal with the application of

violence on others often force its practitioners into situations where more than one motor skill must come into play at the same time, like when moving to cover, searching for targets and reloading a weapon simultaneously. These separate motor skills can be linked through training to other skills. However, this linkage must be relevant and be drilled on repetitively to insure the association. Another well known principle, Hick's law basically states that as the number of response options increases the time the mind takes to select a response increases, thus causing a significant delay in reaction time, possibly as much as 50%. The basic premise of Hick's law has been a long time staple of many in the field of training and has been used as means of justifying a simplification of the training process. It is widely held that if we [instructors] train our students with "too many" tools or options then we may be doing them a grave disservice by slowing down their reaction time. Regardless of your personal take on Hick's law, it does hold some significant validity because it takes time to make a decision, conscious or subconscious, and when we are faced with a series of decisions it takes even longer. However, the condition identified in Hick's law is not 100% constant or even detrimental, as the effects are impacted tremendously on how the decisions are presented, how strong the individual habitually uses the various options and if the options are linked or mapped together via specific stimuli.

Training for simultaneous tasks, like deploying a weapon and a flashlight, are difficult and are limited in the level of performance. However, due to other limiting factors the conditioning for simultaneous tasks is important.

(Courtesy of Perry Taylor)

Generally, even when no physical barriers exist that could keep two or more tasks from being performed simultaneously, limitations in our performance still occur and appear as a decline in our reaction time; this is the essence of Hick's law. In his research, Harold Pashler determined that when a series of options or tasks all begin to compete for execution a mental traffic jam or bottleneck starts, like tour bus gamblers rushing to the free casino buffet line. This is a result of one task waiting for another.

The order in the traffic jam is more or less determined by the perceived priority of the tasks. The response selection process puts the limits of processing at the stage where stimuli are mapped or linked to responses (Pashler, 1994). There are several thoughts on our mental process and how we make decisions and execute responses. Regardless of whether you subscribe to John Boyd's OODA loop, Jeff Gonzales PADE cycle, or any other version, there are generally three steps in our processing function; Identification, Selection, and Execution. Just as we are often called upon to perform or initiate simultaneous tasks, multiple tasks can operate parallel within the identification and execution phases of our processing system. However, all tasks must take turns at the selection phase. Pashler termed the slowing of the reaction process or bottleneck the psychological refractory period or PRP. A number of experiments conducted subsequent to Pashler's original research in to the PRP support his concept (Pashler, 1994). Additional experiments into the human reaction time and decision making process conducted by Anthony Greenwald and Harvey Shulman (1973) clearly showed that some tasks avoid the bottleneck and the buffet line altogether through a system of linking or mapping specific stimulus to responses. This particular process creates a type of connection that is called ideomotor compatible. Additionally, there are some basic physical movements and saccadic movements (tracking movements of the eye) that also bypass delays in processing. The differences in movements and responses that are and are not effected by the reaction time bottleneck are explained through distinguishing between automatic and controlled responses. Controlled responses are those that can be deliberately controlled by the individual and can connect stimulus to a singular response, such as; drawing a weapon, opening a door and applying a control hold. Controlled processes, although capable of connecting stimulus to response, have a limited ability to be linked with other tasks and max the capacity of the decision making process so that when more than one task enters into selection for a response a bottleneck or delay occurs. Automatic processes don't have the capacity limits of controlled responses, allowing for more than one task to be played on at a time. (The driving force that separates controlled and automatic processes in quality habitual practice.) Any task that is practiced to the point of being reflexive can become an automatic process and does not contribute significantly to the delay in our reaction time. However, because this is a subjective theory filled science, just because practice reduces the proportions of interference between responses, it seldom abolishes it completely (Pashler, 1994). Therefore, some interference even with automatic responses is likely to occur, but mitigated tremendously through training. So Hick's law, though not 100% negative, does remain a significant block in developing warriors.

The limitations of Hick's Law can be mitigated or managed through relevant

training aimed at developing automatic responses.

(Courtesy of Perry Taylor)

How do we address the potential implications of Hick's law? Should we remove tools from the student's toolbox and whittle our programs down to one or two simple techniques? It would appear that this is the direction that many would prefer us to go, but I have one question, are all situations the same? Absolutely not! Without a shadow of a doubt no two situations are alike or should be treated alike. Ok, with that out of the way your probably asking, "what do we do?" Well I'm glad you asked, so I will offer some advice.

Abraham Maslow once said, "If the only tool you have is a hammer, every problem looks like a nail." As force application instructors we want to dramatically improve the performance of those in our charge. Training is the single most important tool that we have in this quest. The problem is our training challenges aren't all nails and training alone is not our only tool. Applying sound tools like; course design based on relevant training issues, clear goals and expectations, and developing training environments that help our students reference their experience to real world stimuli are vitally important.

(Courtesy of the Author)

(Courtesy of Mark Burnell)

(Courtesy of the Author)

Realistic, high intensity training works to develop specific responses to specific stimuli based on conditioned experiences.

The environment, regardless of where you are, consists of affordances (terrain, vegetation, and obstacles), which provide the cues necessary for our perception (Gibson, 1977). Additionally, the environment also includes inconsistencies such as lighting variations, shadows, color, shapes and textures that also influence how and what is perceived. It is widely held that perception is a direct end-result of the environment that we occupy (Gibson, 1979) that works parallel with past experiences and conditioning. This mandates that our students learn in realistic environments that provide stimulus that helps to develop perceptual cues that lead to new experiences and compound existing ones. For example consider the problems faced by those who responded to the Columbine tragedy in Littleton, Colorado. This particular incident has been the basis for countless active shooter type training programs and the implementation of "realistic training" for response to these situations, but has this training been truly realistic? Most of the Columbine based training programs teach a series of movement techniques and are often taught in a school type environment, but does this offer the conditioning needed to deal with the challenges of these situations? To some extent, yes, however there is much more going on than simply being able to move down a hallway or through a doorway. In Columbine, a number of environmental conditions existed that should also exist in a training program in order to develop a reliable response. Casualties with visually obvious wounds accompanied by screams for help, shrieking fire alarms and deploying fire sprinkler systems were among the unique environmental conditions that existed in this incident, but are typically omitted in the many "realistic" training programs that abound our industry. Additionally, naturally occurring conditions like; heat, cold, rain and snow also need to be factored into realistic training programs. If an individual, as a condition of his or her occupation, must work in challenging weather conditions then training needs to be conducted under these ambient factors as well. These unique conditions all represent distractions and annoyances that can create delays in making decisions and executing responses; therefore, we must address them in order to develop winners who can work through them.

Training for realism means taking into consideration the actual conditions that one

might have to respond to.

(Courtesy of the author)

When we are dealing with real world, realistic training that takes into account environmental conditions we are trying to instill instantaneous responses to given stimuli regardless of where we are and with the weapons we have at hand. This means that instructors need to condition students to access weapons from whatever configuration they carry them in and within the types of environments where they are likely to use them. This makes no difference if the weapon concerned is a handgun, long gun, sub gun, fixed blade, folding blade, chemical aerosol or baton. The techniques and tactics that a SWAT unit might use in a tight urban setting are different from those that an armed driver would use when seated in a vehicle and again techniques vary dramatically from those that a cop on the street would have to employ. This is not to suggest that we invade our work environments in force and train or new components, but we can certainly reproduce similar environments and conditions during our training sessions.

As far back as the late 1800's research was being conducted that looked at how an individual's thoughts, feelings and actions were influenced by things within their environment that they were not aware they had perceived. A considerable amount of useable information, capable of providing us with decision-making tools, is perceived even though we are not cognizant of the perception, this phenomenon is known as subliminal perception. Over time and through literally 100's of experiments, it has been shown that all types of stimuli can be perceived, even when they are present in forms that make them difficult to tell apart from other types of stimuli. In 1957, James Vicary, claimed that nearly 50,000 customers at a movie theater were exposed to subliminal advertising trying to convince them to eat popcorn and drink Coca-Cola. The advertising was flashed for about 3/1000 of a second, every 5 seconds, while a movie played on the screen. The duration of the advertisement was so brief that the viewer was never consciously aware of perceiving it. However, in spite of not being aware of the message, Vicary claimed that popcorn sales escalated 57% and Coca-Cola sales went up 18%. Since Vicary a number of claims have been made concerning the use of sexually suggestive subliminal words and images in movies and advertising in order to attract would-be customers and sell products. This alleged use of sexually explicit subliminal terms and images also includes allegations about various animated Disney movies. The claims of Vicary and others have never been validated and are generally considered to be false. However, a tremendous amount of subliminal "self-help" products claiming to influence our skills and cure our bad habits are still on the market, even though no solid research exists to prove their claims. What the studies have successfully shown is that subliminal perception, when it happens, impacts an individual's usual interpretations of stimuli, allowing the individual to further reference stimulus and response to a particular environment and event. This perception and referencing could greatly assist in the training and conditioning of our students.

The concept of subliminal perception Suggests that the subconscious mind is able to

acquire information even when the conscious mind is unaware of it.

(Courtesy of SG Five)

Reality Again

Having discussed the importance of a realistic environment, we should

probably discuss what we can do to create this reality. When considering a

realistic training environment, we need to initially set the mental framework

to operate. Most of us in this line of work are normally familiar with the

general use of force model and the force continuum. Personnel outside of

law enforcement and some military circles, may only have a limited

exposure to these concepts, but since we [instructors] will quite likely have exposure to folks with varying backgrounds, it's important for us all to understand it, not agree with, but understand it. The force continuum is, for all intensive purposes, a progressive step concept that illustrates continuous changes in the level of force that can be applied in a given situation. The use of force continuum provides options in a chronological sequence of what should be done first, second, third and ultimately deadly force. The problem is that this concept, or at least how it is most often applied, is not consistent with reality. Often times an individual will enter a situation that requires them to apply deadly force without considering lower levels of force. In reality situations can escalate and de-escalate, stop and start and change directions without warning. An understanding of the nature of high-risk situations is absolutely critical to developing not only appropriate responses, but also realistic environments.

A more accurate interpretation of the force continuum would be to consider it a guide to a series of options that are available for the appropriate control of a given situation. The individual should be trained to exploit options and understand that control of the situation is the ultimate goal. Further, the individual must be trained to recognize when compliance or control is obtained and prevent both over-reaction and under-reaction. In relating the force continuum to developing our environment consider this, the continuum is based on options, all environments present options, and options are tools for the toolbox.

Environmental options that correlate with a use of force continuum can be things like; obstacles and barriers, avenues of egress and escape and potential weapons.

Conditioning for response includes taking the time to position obstacles between yourself and the potential area of threat. Here an officer leaves the intermediate barrier of a car door and the distance to the suspect vehicle as barriers.

(Courtesy of Perry Taylor)

Obstacles and barriers represent both impediments and control tools. The quick and decisive use of a physical object to block an individual's path or route them in a certain direction can have a great impact on our subsequent actions. Likewise an object in our path can cause a delay in our ability, also impacting our subsequent decisions. Avenues of egress and escape provide our student or his foe with a means to run and live to fight another day. Knowing the location of escape routes and safe locations to fall back to is without a doubt a critical component of tactical operations (SWAT, Hostage Rescue, etc.), but is a seldom-considered component of individual skills. What represents an egress location or an escape route? That's easy; doors, windows, hallways, tunnels, crawl spaces, the gap between two parked cars, the space under a table and the possibilities go on, literally an endless list. An awareness of potential weapons within our given environment is a seemingly simple consideration. An obvious array of unconventional weapons like; kitchen tools, sports equipment, and gardening equipment exist in nearly all homes in our country. However, what about considering those things that exist well outside the box and far past unconventional? What am I referring to you ask? Quite simply, I am referring to literally anything and everything we can lay our hands on! Brooms, silverware, table legs, lamps, heavy books, potted plants, handful of dirt and anything within your immediate environment that can be considered for use as a weapon, often quite effective weapons. This portion of this chapter is undoubtedly going to be met with controversy by many who read it, as a tremendous number of organizations not only train, but craft policy that prevents and discourages the very idea of striking another human with anything not "authorized" by the agency. These preventive methods are typically backed up with the potential of severe discipline in the event that they are violated. Even if the agency has no intention of enforcing the policy, it sends a dangerous message to those it applies to. In my early days of Army life during my first tour in Korea, I and 200 or so other GI's sat through hours of in-processing briefings. These briefings were hard enough to follow, but coming on the heels of a 16-hour non-stop flight made them all that much more difficult to bear. One always seems to stick out in the minds of young male soldiers and that is the one about catching VD and getting kicked out of the Army. Now this was more scare tactic than it was reality because most commanders never knew if one of their guys got VD or sought medical attention unless it went untreated. Unfortunately, the message that was sent told the young soldier that he was in very deep trouble if he got VD. So when it happened, many soldiers would not go and get treatment, thus developing into more serious afflictions, so goes the message of bottomless policy. When a situation degrades into terminal circumstances all bets are off and those who we train must be made aware of it. An additional benefit of an awareness of potential weapons in an environment is in controlling the opposing subject or subjects. Losing control of a subject and permitting access to anything that can be used, as a weapon can be at the very least, disastrous. While our students may be held to standards of conduct and ethical rules prohibiting cheating in a fight, their opponents are not. Additionally, since their opponents have their very freedom at stake they can tend to be very motivated in their quest to inflict injury. Combine this deep motivation with the lack of restrictive rules and we have some significant disaster multipliers.

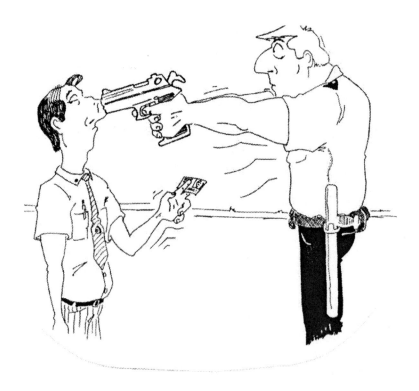

Conditioning for a timely and appropriate response must also include conditioning for situational awareness and appropriate weapons selection.

(Illustration courtesy of Mark Burnell)

Static v. Dynamic Training

In nearly all fields of our existence, innovative technology and concepts have dominated and set forth the standards. Regardless of the industry, those involved have had to change and shift with this technology in order to compete. As new and sometimes better innovations and inventions are thrust into the world we are forced to learn and use them. This is more or less a logical and sequential process, occurring since man became man. However, the field of force application as a whole is very resistant to this progressive change and is evident in the history and case law that has been handed down over the years. The standard doctrine for many is to stand, static, in a location and work physical techniques; like, baton strikes, control holds, handcuffing or other empty hand techniques. This is also true and typically more evident for those who train with firearms. Understandably, this concept of finding and holding a static position to train from, particularly with firearms, comes from the earliest beginnings of organized training where we took many things from popular sport and competition circles of that era. The problem is that our world has developed into a domain that is anything but static and safety and control are usually best found through a solid application of movement.

Only a limited number situations are truly Static. The essence of tactics is the ability

to move from position to position and to effectively close on your objective.

(Courtesy of Perry Taylor)

I will stop here for a moment and digress a bit as I don't want to be misunderstood anymore than I may all ready be. Static training has a definite place in the training environment and when used properly and applied like a stepping-stone to more advanced training it works great. For example, having student work set distances from static positions on a firing range while they receive instruction on marksmanship or weapons handling is a great way to maintain control and safety of a class. Further, it allows an instructor to judge the ability and determine the progress being made. The problem is, many who teach never break from the static training environment, preferring to work constantly on square ranges and at set distances. When this occurs the student is handicapped because he or she never gains any more experience than what that square static range environment allows. A classic example of this is the old and ingrained habit of firearms instructors referring to a safe direction as "down range". While a shooting range obviously has a down range area, where in our real world is "down range"? When a shooting occurs on the street, in the desert, the jungle or in an aircraft there are generally no specially pre-designated "safe" or "down range" area, but students are constantly being told to keep the weapon pointed down range. What about changing the terminology to something else, keep the weapon pointed towards the area of threat or simply having them keep it pointed in a safe direction? Additionally, it is a popular concept to have shooters on line learn to scan left and right for additional targets prior to holstering their weapon, but the weapon must remain pointed "down range" during this procedure. We teach our students to operate, hopefully we do, in real world environments where we trust them to carry and use firearms. It is reasonable to assume that at some point in their career the firearm will have to come out of the holster or the vehicle rack and be deployed into the real world. We trust these individuals, no, we depend on these individuals to do this when necessary, but we treat them like children when they are on a range. I understand the issue of safety and the fact that inappropriate and unnecessary handling of weapons can lead to disaster. However, I firmly believe that to beat this "disastrous" condition we need to train our student to handle them appropriately in environments where there is no "down range"!

On ranges we have clear defined down range areas, but where does range exist in the real world?

(Courtesy of the author)

There are many techniques within the training community that emphasize safe and controlled movement with a deployed firearm. Among the turning and pivoting techniques that all emphasis safe, yet tactically efficient movements, I have found one to be particularly simple to teach and effective for all levels of students, this being the safety circle. The National Rifle Association's Law Enforcement Activities Division or LEAD is the organization that teaches this technique. Having countless years of active interface with an astounding number of law enforcement and military agencies, LEAD has developed a significant program based on real world practical handling of firearms. I'm unsure of the origin of the safety circle, but I have found it simple, effective and applicable to all forms of firearms. Imagine a circle on the ground that goes around your entire body with a circumference that extends past your feet by about 6-8 inches. The area inside the circle is your personal domain and the area outside the circle is the open domain of the real world, both of these domains are considered neutral unless a threat is present. The circle itself, all 360 degrees of it, is the safe zone. As a shooter finishes negotiating the targets in his or her specific area of threat and scans to the immediate left and right of that area for additional threats, he or she needs to check the remaining area of their environment for threats including the area behind them. The shooter, with finger off the trigger and straight along the frame of the weapon, drops the weapon to a position where the muzzle is pointed directly at the imaginary circle on the ground. The shooter can pivot or turn within this circle and allow the weapon to rotate along the imaginary circle. Obviously there are those out there who are saying, "why not just look over your shoulders?" and certainly you can. However, the concept of this book is winning and in that concept it is my contention that it is simpler and more effective to engage a threat if I am physically oriented to it with my body and both of my eyes, rather than rely on my peripheral vision and the time it would take to re-orient myself after I realized it was a threat. Now I suppose there are probably a few of you out there who can shoot over your shoulder and between your legs, using only your wife's compact but I cannot, nor is it reasonable to assume that the majority of our students can either.

The "safety circle" (illustrated with the aid of a child's toy supplied by the authors youngest daughter), represents a safe method of having a weapon deployed in a real world environment.

(Courtesy of the author)

At face value, the safety circle does not appear too dynamic, at least by most people's impression of the term dynamic. Most of our students, being involved in some form of law enforcement or military service, envision men in black crashing through a door and shooting everything in sight when they hear the word dynamic. However, dynamic when used in contrast to static refers to training or actions that incorporate some level of reasonable and relevant movement. Certainly, dynamic training can progress to very high speed running and shooting exercises but as stated earlier in this text, we must walk before we run and we must run before we jump from the plane. The safety circle is a small example of moving to a simpler form of dynamic training and is not intended to represent the entire gambit of dynamic training. It is only intended to get you the instructor to think. My point is that while static training has its place in force application training, to truly condition an individual to respond appropriately and function with a winning attitude we [instructors] have got to move forward in our training and incorporate some dynamic activity that involves movement.

Principals of Movement

While there are certainly many different forms and concepts of movement the basic idea behind this book was that it was going to be used by professional trainers dealing with folks who move and act with tactical efficiency. So when discussing movement, I will focus primarily on movement that is associated with the use of tactics. The ability to move is the essence of most tactics. Without movement, we are left only with the tactic of standing still and, although of potential use, is not much of a tactic. Essentially, we move tactically, dynamic or deliberate, for the purpose of covering an open space to either close on a threat or in egress, to open the space between you and a threat. It is without doubt that when we move we must move with balance and stability attempting to maintain a solid platform that we can either shoot or fight from. To maintain this solid, stability we need to move in a manner that avoids exaggerated actions and we must keep an awareness of our environment to prevent unintentional collisions or accidents when we move. We all have experience walking and have been doing it with some success for a long time. However, coordinated movement on real world topography creates some complications. So for our guys to win, we have to apply some serious thought to movement.

Forward and back movement is the first and most elemental direction of movement that we will discuss. In my classes, because of my training and background, I teach a stance and movement technique that is consistent with weapons use, physical tactics and tactical movement. I teach it this way not because of my fear or belief of Hick's law, but because it's simple, effective and consistent with typical human reactions under stress. The position is similar to the infamous "FI" stance with one exception, instead of blading myself to protect my sidearm, I turn so my upper torso and hips are squared off towards my area of threat or direction of movement. Generally, my feet are about shoulder width apart and my primary leg is to the rear about a half step, the knees are bent slightly. During forward movement, it is important to walk heel to toe in order to allow the foot to literally roll over obstacles or debris that could be in our path and the opposite, toe to heel when move to the rear. From this position, I can face an attacker in a physical clash, shoulder a weapon and fire it with good accuracy and recoil control and most importantly, I can move with stability and control. Is this stance any different from ones taught by other instructors? No, it's not. This is a very simple and commonly used technique that is seldom taught or reinforced to students even though it allows for a tremendous amount of control.

The "forward stance" is a very simple technique that is found in many popular

forms of Contemporary martial arts. The stance is adaptable to many forms of

armed and unarmed combat, as well as tactically efficient movement.

(Courtesy of the author)

Another view of the "forward stance" illustrates how the stance places weight

forward and squares the individual off towards the area of threat.

(Courtesy of the author)

Lateral movement from this position can be facilitated two ways. First,

we can move side to side by simply opening our stance and closing it. This

is a very simple, yet effective version of side stepping that keeps us in the same posture and allows us to maintain that solid stable platform. Second, and this comes with some difficulty, we can side step by crossing our legs. Generally, crossing the legs is frowned upon because at the moment when the legs are crossed we become grossly unbalanced and vulnerable to being knocked down. Further, it's difficult to break into a run when our legs are crossed, thus diminishing our mobility. However, if practiced this is a very comfortable and stable technique that is used very effectively by those who invest time in working with it. An added benefit of walking and crossing the legs during stealth movement is that you have a tendency to roll the foot at the side with more control and less noise than when stepping on to the ball of the foot. When walking and crossing the legs it is important to remember to step and cross to the rear or behind, as this method keeps one foot in firm contact with the ground as the other one is in mid-stride. If we step to the front, the tendency is for the base foot to cant on its edge in the direction of travel while the other foot is in mid-stride, creating a severe balance problem. Again, crossing the legs can be effective, but only for those who practice it and develop proficiency; our typical audience is filled with those who won't invest the time to practice so it is probably best to avoid this technique and focus on more simple and effective techniques. Noise and light discipline are two of the most commonly overlooked facets of movement, particularly in law enforcement circles, and are typically dismissed even by those who are aware of them. However, they are extremely important even when the movement is not stealth in manner. Keep in mind that any training, especially force application, is a series of applied discipline otherwise we're nothing more than bulls in a china shop. An awareness of our environment and an understanding of noise and light discipline are crucial to positive and aggressive movement. For example, we may not be moving with stealth as our purpose, but we still need to remain cognizant of loose footing or rubbing our gear and equipment against walls (remember the prior section on environmental training?). In stealth movement, for obvious reasons, we need to apply the principle of noise and light discipline since we want to remain undetected. However, one only needs to sit back and watch most grown adults lumber around the world to realize that most folks aren't really aware of the amount of noise they make when they move. Watch your students in training and see how clumsy they are. Several years ago, I and a number of other instructors at the Sheriffs Office where I worked put together a reduced light building search class as a portion of the agencies, annual training program. We were given a large dormitory at the agencies adult shock incarceration center as our training area. We had no live roll players, no live ammunition and no moving targets or manikins. What we did do was move the bunks, fill the room with piles of trash and other debris and turn the lights out. The officers were to move as a two-man team to search the structure for a suspect that had fled the scene of a crime. The officers had their duty sidearms and flashlights. Literally at the moment of entry, the majority of the officers began to hyperventilate (insecure with the unknown). The movement became awkward and clumsy as the officers literally crashed into every bunk and stepped into each pile of debris. Each team moved noisier than the last, but during the debriefing not one team could remember creating the noise. The fact that the officers had grown careless in the movement patterns and comfortable in the amount of noise that they made created a serious deficit in their training and a hazard should they have to move with stealth.

Our primary responses in most force application situations come down to offense and defense or our ability to attack and counterattack. These are the basis for all combative applications, be they beginner or advanced. The application of any offensive or defensive act requires some form of movement and some special considerations to maximize that movement. This applies to the simple front, back and lateral movement already discussed and to that detailed movement that is done up close and personal when fighting with another human being. When our distance is extremely close and we have consciously chosen to respond with a specific tactic or technique, our ultimate goal is generally to neutralize and control an opponent as quickly as possible. To effectively accomplish this, we have to focus on four elements when we move. These elements were identified and termed by a Korean gentleman named Hanho (1992) and they are as follows:

1. Time, every movement requires time to initiate and execute. Always keep in mind that action is faster than reaction.

2. Stability, most dynamic actions cause us to lose our balance and stability for at least a brief period. When we walk and run, there exists a brief period of time when our feet leave the ground entirely. All our movement and subsequent techniques must be developed on a stable platform just like shooting.

3. Superiority, by developing our responses through various psychological and physical training methods, we can gain a marked advantage or superiority over our opponents.

4. Opening, through the use of maneuvering and physical tactics, we need to condition our students to look for openings in their opponent's posture. These openings give us the opportunity to apply appropriate and effective techniques and, most importantly, let us determine when we have gained control or our opponent has complied.

The elements above apply equally to all force application situations, empty hand physical tactics and firearms application situations as well because all purposeful movement requires practice and careful consideration. When we concern ourselves with developing a warrior into a winning warrior, we must emphasize the basis of movement and how that movement relates to tactics and strategy. When an opponent is trained and conditioned in a superior manner, an untrained or ill-prepared opponent will have a difficult time defeating him. The individual who possesses the greater skills and ability to use strategy will have a dramatic advantage over the other. Professional boxers spend countless hours developing their footwork, timing and movement skills working far past simply their ability to punch. As a boxer moves about the ring, he is constantly searching for openings in his opponents techniques that will allow him to dominate the fight and ultimately win the bout. The practice that goes into movement depends heavily on the mindset of the individual concerned (sound familiar?) as does a tremendous portion of force application training; unfortunately, this mindset issue is seldom passed on from teacher to student. This situation becomes painfully evident when an individual is not held accountable for using the training that they have received when they return to their parent agency. It's common for agency administrators to view training as some form of necessary game instead of an imperative part of the occupation that mitigates risk and liability.

Reduced Light Conditions

The FBI and the Department of Justice tell us that a vast portion of a police officer's day occurs during periods of reduced light or diminished visibility. This statistic is based on the idea that two thirds of the day, the earth is in a position to receive reduced sunlight. However, a police officer typically encounters many forms of reduced light through the full 24 hours of a day. Situations like searching the interior of a vehicle or the darkened inner sanctum of a home used as a methamphetamine lab are realistic representations of just a few situations where reduced light is encountered during a "normal" day. A paradox occurs in this situation that makes little to no sense. As I travel about the world teaching various courses in weapons and tactics, I often ask my students, "Who carries a flashlight when they work day shift?" Disturbingly a number of students in every class respond with "I don't need it, I'm on day shift" or "It's home in the charger". A tool as simple as a flashlight is so easily and so often dismissed as an unnecessary item I find this to be disturbing especially as often as it is used.

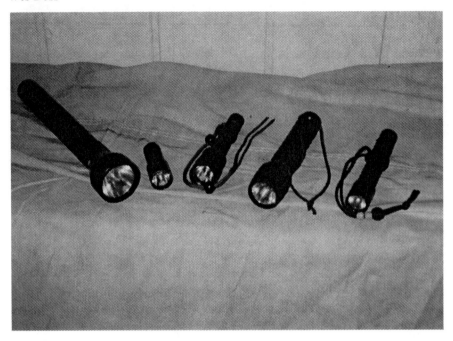

Non-dedicated illumination devices (flashlights) come in all shapes, sizes and power out puts. Many are now designed specifically for tactical applications and the rigors of out right combat.

(Courtesy of the author)

Among the five senses, sight is our primary sensory contrivance. Turn out the lights and reduce our ability to see and our other senses try to compensate, but we still remain heavily dependent on our sight. As the input becomes impaired, the mind augments the limited information with assumptions about what can't be seen. This presumptuous action is based on each individual's prior training, conditioning and experiences. If an individual's experiences are all founded on fears or on irrelevant training, then the gaps in the sensory input stand a strong chance of being filled with heightened fearfulness and an inability to mitigate the situation. If you doubt the power of the human mind to create environmental assumptions consider this, in the 1970's the movie *Jaws* played across the country. This one movie, the first of many great disaster type movies, sent masses fleeing from all bodies of water at the mere mention of sharks. This extreme reaction extended far beyond the open sea, as it was not uncommon to see people afraid to swim in lakes, rivers and even swimming pools. The point is the human brain is both a wonderful thing and a horrible beast, as it allows us to be consumed with false senses and assumptions.

In an attempt to address the obvious but often ignored problems with reduced light situations a number of popular techniques have been developed to facilitate the use of a weapon and a flashlight. Each of these techniques has received tremendous acclaim and bears someone's name or catchy moniker, but does very little to address the legitimate use of a flashlight in both day-to-day operations and those unexpected situations involving the use of force. Each of the popular techniques function under the presumption that if a flashlight is to be used, it must be used in concert with a weapon and all search activities must be led with the weapon's muzzle. Further, these techniques begin and end from a "Ready" position, with both flashlight and weapon deployed. However, the majority of search situations necessitating the use of a flashlight seldom require the deployment of a weapon. In fact, most police officers and other professionals, engage in casual non-threatening use of a flashlight throughout the day, but are typically only provided with flashlight "Shooting" techniques. What actions will an individual take when a "routine" situations goes terminal? What options will the individual have if he or she has not been trained to transition to an aggressive posture-deploying flashlight and weapon simultaneously? Consider the infamous recovered unit video from the Texas constable who, while conducting a traffic stop, was attacked and killed. During this stop the officer allows the three occupants of the vehicle to get out and surround him while his attention is focused on the trunk of the vehicle and its contents. During this incident, from the start to the grisly end, the constable is holding his multi-cell flashlight. The constable tries to fend off his attackers, but never uses the light as a weapon or lets go of it. Typically, reduced light training consists of individuals being shown various methods of operating a flashlight and a weapon together. The individual is then told to select a method that he or she is comfortable with. This "training" is conducted on a range at defined distances and from a static position. The dilemma is that most lethal force situations are not static but are extremely dynamic and unless an individual has already assumed a position with a weapon and flashlight deployed, he or she is not likely to use the flashlight regardless of the level of light present. To compound an already monumental problem, the methods taught are all solely target engagement techniques and do nothing to address drawing, searching or moving with a flashlight. These techniques also fail to address physical control of the weapon should an opponent move within the individual's reactionary distance or actually engage the individual in a physical struggle. Some might believe that an individual may be better off not using a flashlight and certainly under some circumstances this is true. However, the safety of the public, our team members, our clients or our family is a responsibility and in exercising that responsibility one must positively distinguish between threats and non-threats. Because target identification under reduced light is difficult, we need to use flashlights as much as possible and their use needs to be a conditioned, reflexive action. I, like many of my contemporaries, still teach the various techniques for using a flashlight and a weapon. I have been guilty for many years, and now this makes me sound like a hypocrite, of only teaching the target engagement portion of reduced light situations. However, while I still teach what works, I also adamantly preach flexibility and repetitive use of the flashlight from unconventional positions and unconventional situations. For example, after some serious dry fire training, I may have a student conduct an evidence search of a vehicle and at some point he is confronted by a potential threat that he must identify before deciding what to do. This scenario, although

177

simple, forces the student to come from a position of disadvantage and utilize the flashlight in a decision-making capacity that may or may not require the use of the weapon. I also run reduced light range training from seated, prone and supine positions under conditions where the student may have multiple obscured targets before him that must be identified before he decides to shoot.

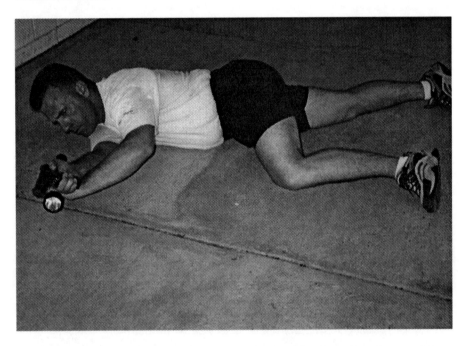

Teaching the use of a flashlight from a Standing position is really only scratching the surface of dim-light training. A great many situations end up on the ground, thus justifying the need for ground based training.

(Courtesy of the author)

Flat on the back is a compromising and Difficult position to fight from, particular attention should be paid to such unconventional positions.

(Courtesy of the author)

The first step, regardless of the task or the threat level, is to get individuals to move the flashlight out of dominant hand and condition them to use it in the non-dominate or reactionary hand. This seems very elementary, but one only needs to watch people at work to realize that the light always ends up back in the dominant hand. Naturally, humans rely on their dominant appendages for mechanical tasks but by not breaking the reliance on the dominant hand, we encourage both a physical and mental barrier to accessing the weapon when we need it. This condition is consistent with both Pashler's bottleneck PRP theory and Hick's law. When

179

we train our students to win in the reduced light arena, we need to work them under four specific types of scenario's involving potential threats: with only the weapon deployed, with only the flashlight deployed, with both flashlight and weapon deployed and with neither weapon nor flashlight deployed. However, most individuals are only provided training in limited doses dealing with potential threats with both weapon and flashlight deployed. This is quite antithetical to our role as instructors who are trying to adequately prepare folks for real-world situations.

As previously discussed, as no doubt I will bring up again, typical training seldom reflects the limitations of duty equipment or the complications that the environment itself can create. Body armor, duty uniforms, leather gear and military load bearing equipment do restrict movement and eliminate options that individuals can grow comfortable and confident in. Furthermore, elements like narrow hallways, cluttered rooms, oily floors and uncooperative individuals create a significant impact on strategy and tactics. Thus, all training should be conducted as realistically as possible. The previously stated scenarios for flashlight/weapon training are without a doubt inconvenient and because they require ambidextrous activity, they are difficult to administer. However, they are not impossible and they are positively essential for reliable work in a reduced light environment.

Often uncomfortable and limiting in its Design, duty gear should be worn if in fact
it is worn during the course of an individuals work day.

(Courtesy of Perry Taylor)

Any training having to do with an individual's skills at situational control, especially operating in a reduced light environment, is critical for operational success. When lives are at stake, there is far too much to lose if training does not address actual issues or is not done with ample intensity. How we train and what we train our students to do is critical. The people we train cannot afford to make a mistake when a confrontation occurs, whether it is light or dark.

Chapter Seven

"Though defensive violence will always be a sad necessity in the eyes of men of principle, It would be still more unfortunate if wrongdoers should dominate just men"

-St. Augustine

Weapons and Physical Tactics Training

As previously discussed, in the macrocosm of force application training there exist two distinctly different worlds, the subjective and the objective. The subjective world being that which is influenced by the individual's perception through stories and visions supplied by various forms of popular media, suggesting that excessive violence is a way of life and the good guy always walks away unscathed. The objective world is again that domain that is based on real life and brutally harsh realities of our various occupations. A problem exists in the objective world that creates immense complications into what is already a complicated environment. These complications stem from an unusual and often-antagonistic relationship between those involved in firearms training and those who teach defensive tactics or unarmed self-defense. Although there are other conflicting situations that create complications, firearms and physical tactics are the primary components of what is generally considered use of force training.

Historically, the elements stem from distinctly different philosophical origins. It is these conflicting, possibly competing, philosophies that causes a dangerous breakdown in the relationship between two vital ingredients that should work in unison, but typically don't. It is my purpose in this section to address this common defalcation and to offer a modified approach to conducting force application training.

In its conceptual form, force application training seems relatively focused and need oriented, but upon application the situation can change dramatically. Take the analogy I used of the child's balloon. As it became stretched to it's maximum capacity the unnoticed flaw is breached and the balloon ruptures. This is a great analogy to describe the friction that often occurs between firearms and defensive tactics (physical skills) training that rises to maximum capacity unnoticed until it's too late. Firearms and physical tactics are primary components of what we considered use of force training; both is oriented towards individual safety and control skills. Unfortunately, the physical and mental conditioning applied to each is so radically different that, under periods of high stress, they can conflict with each other. It is this conflict, like Pashler's PRP bottleneck, that can render both elements less than effective and create an individual safety nightmare.

I know I've said this before and trust me I'll say it again but to start with, instructors need to know and understand how humans learn. Although there are many schools of thought concerning learning and educational psychology, learning can be generally summed up as the process of gaining knowledge or skills. Coupled with the memories ability to store and retrieve information, learning helps us grow and develop. As a species, human beings learn from the moment of birth, maybe sooner, and extend the learning process far past school and childhood experiences; essentially we continue to learn throughout our existence. Learning generally takes place in one of more of three domains; cognitive, psychomotor and affective.

The cognitive domain is entrenched in the mind's ability to purchase, preserve and regain knowledge that is built on previously established knowledge or experiences. Typically information taken in that affects the cognitive domain is done through lecture and when in an academic environment, is usually evaluated through both subjective and objective testing procedures. The affective domain finds its foundation in human behavior or our individual beliefs and values. In the affective domain we examine, asses and amalgamate information, forming our own ideas and assumptions based on what we have experienced. Information on this domain is typically transferred through debate, argument and discussion. Evaluation is usually completed through the use of some sort of checklist to ensure that relevant or critical points are covered. Evaluation of this level provides no clear right or wrong positions, since the area is based on the beliefs of the individual. The psychomotor domain is a skill-based domain and while all training effects the various learning domains in one way or another. We find ourselves dealing with the psychomotor the most often. This domain is stimulated through the student observing a demonstrated activity conducted by the instructor and then by repeating or imitating the activity. Further, the student will continue on to greater stimulation through repeated practice until the action is committed to habit. Typically an individual is evaluated on this domain through a performance or skill test. For obvious reasons the cognitive and the psychomotor appear somewhat synonymous, but if the activity requires a physical application of a skill and requires more resources than what a classroom can offer it is probably stimulating the psychomotor domain. Now that I have once again re-hashed the learning process lets get back to the conflict we were discussing earlier.

In firearms training, we work to develop a strong, stable, static platform that allows the body to stabilize. This stationary stabilization, like a tripod for a camera or a crew served weapon, is necessary for slow accurate shooting. In fact, stationary stabilization is a principal constituent of effective muscle control that humans rely on, subconsciously, to sit, stand, walk and generally remain upright. However, for martial arts type moves and techniques, like those popularized is many defensive tactics training programs, to be effective requires a dynamic form of stabilization based on movement and weight transfer to establish stability. It is this difference in stabilization needs that initially separates firearms training from defensive tactics training. This separation grows as stationary stabilization finds continuous reinforcement through many daily activities like; sitting at a desk, driving a car, using a computer, and writing a report, does this sound like anyone's average day?

Typical firearms training works to develop a strong, stable and static platform allowing for slow, accurate shooting.

(Courtesy of Perry Taylor)

In contrast to typical firearms training, the reality of situations demands stabilization be found through dynamic movement. Real world shooting situations don't always happen at set distances or on flat ground.

(Courtesy of the author)

It is well documented that humans, when faced with extreme stress stimuli like that, which occurs in a violent encounter, revert to a response based on training or conditioning and a sense of confidence in that response. If the sense of confidence is absent, the mind will select a different response. Many individuals act in response to a high-risk situation, particularly close quarters situations, by dropping into a solid, stationary, stabilizing posture. The stances most typically assumed are some semblance of shooting stances or the much maligned but ever popular "FI" stance. Although having no intention of drawing a firearm or taking a statement, it is also quite common to adopt the footwork and body posturing of shooting stance in preparation of an oncoming physical conflict. Right or wrong, the individual is simply reacting in accordance with training received and reinforced countless times. The problem lays in the fact that these stances by their very nature are not conducive to the application of physical tactics, be it punching, kicking, control holds, takedowns,or the use of a baton. Most techniques, especially control holds and take downs, are designed to be applied with little effort on the part of the he who applies them. These techniques generally use momentum or exploit an opponent's lack of balance for effectiveness. From a traditional "FI" stance or popular Weaver or Isosceles stance it is difficult, if not impossible, to execute the steps and movements needed to apply the techniques. Forcing the individual into a situation where he or she is using valuable strength and typically this is the point where these situations become terminal. It is extremely difficult to generate and control the force required to subdue a dedicated and actively resisting opponent, causing a great many physical confrontations to turn into unintentional wrestling matches where someone frequently gets hurt or killed unnecessarily (MacYoung, 1999). It only takes one of these "wrestling matches" to realize the deficiencies of many approved and heavily endorsed programs.

The traditional "FI" stance, though popular and consistent with a number of popular shooting techniques, is a difficult position to apply unarmed techniques from, should a situation degrade to a hand-to-hand conflict.

(Courtesy of SG Five)

189

As armed professionals, we typically are in positions necessitating us to react to an action that has already occurred, starting us from a position of relative disadvantage. This will remain the norm until such time as both law enforcement and military in our country begin the practice of preemptive strikes against our opponents. Being reactive we further complicate the situation by operating training programs that are all predicated on the assumption that lethal encounters are nothing more than a distended simulation filled with unrealistic adversaries that don't act or respond as an actual combatant would.

Additionally, our mental state of readiness or preparation and our physical positioning, along with movement skills, create the base from which all our fighting abilities are built on. Conditioning students to adopt a set of techniques that can be used in a multitude of situations, physical or weapons oriented is of paramount importance. Without an effective base of skills, most of the "tools" taught cannot be executed to their full potential. Finding or teaching a flexible and adaptable concept requires the coalescence or fusion of all use of force skills. Until this unification takes place, we will continue to face tremendous and ongoing potential for disaster.

The limited time and distance involved in close quarters engagements necessitates an understanding of transitioning from firearms to other weapons and techniques.

(Illustration courtesy of Mark Burnell)

As trivial as it sounds, the stance, footwork and mobility skills that one either possesses or has developed truly effects ones overall capabilities and has a serious influence on the outcome of a situation. Our positioning must allow for a solid shooting platform at the same time giving us the ability to remain explosively mobile. General guidelines, not commandments that are set in stone, for the use and development of a reliable control position are simple but typically neglected. First of all, a trained and prepared professional must strive to remain relaxed and breathe. I fully understand that to remain relaxed in the middle of a desperate situation is extremely difficult. However, through training and conditioning it is a very realistic concept. It would amaze most just how many individuals actually cease breathing when under intense anxiety. It's critically important for us to remember that our various senses are the instruments through which we receive information. During a high stress situation, especially one that is perceived as uncontrollable, some of our senses literally shut down and others grow stronger. It's very common to experience limited hearing and loss of fine motor skills, while our vision becomes more acute and focused. The reliability of these various senses is heavily dependent on the body's ability to maintain a steady supply of richly oxygenated blood. If we stop breathing, even for a little while, all the shooting and punching skills in the world won't do us very much good.

The general stance that I teach and operate from as a base-line position is essentially a modified boxer's stance and is the very same stance that I mentioned in the previous chapter. Although it is known by many names, for simplicity, I have always referred to it as a forward stance. The upper body is squared at the hips towards the direction of the perceived threat. The arms can be down to the side or held in front of the body with the elbows down and the hands at about chest or face height for protection. The legs are separated at approximately shoulder width with the reactionary leg forward of the body slightly and the primary leg slightly to the rear, in a posture similar to the one taken when pushing a car or a large piece of furniture. The distance at which we place the reactionary and primary legs can be varied depending on the needs of the situation. The knees are never locked, instead the legs are kept slightly bent with the body weight kept forward in order to absorb the shock of an incoming assault or ride the recoil of a weapon. Keeping the knees slightly bent also allows us to maintain a posture that permits quick movement in all directions. Compare this position to both the standard "FI" stance and a normal standing position and try to move quickly, present a weapon or fend off an attack. In doing this comparison, one will quickly find that it is difficult to keep your balance or move fluidly when utilizing one of the age-old techniques, but just how easy it is to maneuver from the more aggressive forward stance. The forward stance is an "open" stance; similar in function to an open batting stance used in baseball and with similar attributes. Like an open batting stance, the forward stance places an individual into a position that focuses attention and physical power towards the area of threat and allows for greater visibility over the area of concern. This greater visibility allows for more information to be taken in initially and as the situation progresses and vision becomes more focused due to SNS activation it places an individual into a status where one can utilize this change in visual reception to its best advantage. Unlike the bladed and closed "FI" stance that not only puts your area of threat into a sector of your peripheral vision, but takes the body off line and into a position of disadvantage when having to meet force with force, like during a physical assault. The beauty of this stance is that it is not a muscle intensive technique, yet is extremely powerful and it easily dovetails into other areas of force application training. Now, I'll stop here and enlighten you a bit as to my reasoning behind using the forward stance as such an all-inclusive example. By no means do I mean to imply that this stance alone is the sole method for bridging the gap between these two concepts. I simply mean to use it as an example, like the safety circles, of how we can alter our training programs to be more relevant to the actual needs of our students. The shear fact that firearms training and physical tactics are typically on opposite ends of the spectrum is a most unsatisfactory situation. In the real world it is quite likely that a police officer, a soldier or a protection specialist will have to use both systems, contemporaneously, in any number of conceivable situations. However, if we don't prepare our students to employ both elements simultaneously, we can't expect them to, thus leaving them further behind the power curve and in a position that is not conducive to winning.

The "forward stance" is applicable to force on force situations as well as weapons related situations. Even when an officer deploys into a location with a weapon he must be ready to transition to other techniques should he lose the weapon or it become disabled.

(Courtesy of Perry Taylor)

Additional Problems

Typically, at least in most organizations, there stands a set of double standards when it comes to the use of force. This is primarily aimed at law enforcement but could apply to situations faced by entities within the military. While specialized units, like SWAT teams, are usually afforded the latitude to train for any possible encounter, patrol forces are not. Again, falling on the heads of an agency's administration, patrol officers face a barrage of negative reactions when they train, or attempt to train, in "unconventional" techniques. By unconventional I simply mean anything outside of the norm established by the academy. The mere suggestion that officers be trained to employ things like head butts, elbow strikes, eye gouges and the unholy politically incorrect knife, often brings fits of disapproval from the powers to be within an agency. Personally, I find this remarkably odd, since we train them to shoot out the upper body cavity of other human beings. Unconventional training is most often left to the realm of the SWAT cop, where they are ordinarily granted the latitude to train with the idea of "do what ever it takes". The latitude given these special units represents what is right within the police training arena, worst-case scenario preparation, on-going resource familiarization, teamwork, tactical awareness, reality-based training, departmental and public support are just some of the reasons why these officers are able to overcome such adversity

(Fourkiller, 2002). Tactical officers are afforded the liberty to train, but in reality tactical operations account for a very small amount of the violent assaults encountered by officers every year. With this is mind, one would think that the officer on the street would be afforded the opportunity to train harder and with greater realism. The unpredictable actions and violent tendencies of offenders and the uncertainty faced on many "routine" calls often equates to violent assaults against police officers with the potential of serious injury and far to often, death. With this in mind, it would be reasonable to assume that all officers would be afforded ongoing training and resources that would help to win these encounters, unfortunately this is not typically the case.

For several years' only tactical officers were afforded training in unconventional

techniques. However, it is the street officer who has a greater likelihood of

spontaneous contact and conflict.

(Courtesy of Perry Taylor

The defensive tactics training received by most officers in this country,

understandably there are exceptions, is generally inadequate, unrealistic and

in many cases counter productive. The problem is, as with many martial arts

schools teaching "self defense", that the majority of techniques are geared

towards the domination and control of passive or minimally resistant

opponents. Additionally, a great many of these systems also rely on the

notion that an additional or back up officer will be there to assist. This is so far from reality that I can't imagine where they draw these concepts from. A tremendous percentage of law enforcement agencies in the United States are rural and an equally high number of agencies have 100 officers or less. I worked for just over ten years for a rural county sheriff's office in Arizona where my additional officer, my back up, was thirty minutes or more away. Sometimes my back up had to come from a sister agency that was even farther away. The whole world can change in well under 30 minutes and most who will assault police officer won't wait for the reinforcements to arrive. Most "systems" provide pain compliance techniques, a limited number of strikes, a limited grasp of weapons retention and some very basic grappling techniques. These systems tend to be packaged in 3 to 5 day blocks, at the end of which the student becomes a "master". To divagate back to a previous adage in this section about the often diametrical opposition of firearms training and physical tactics training another such deviation exists in the form of defensive tactics that are taught. In basic firearms training, we train for stoppages and malfunctions, regardless of the weapon system. We train to fix a problem and to resume the fight as soon as possible. After all weapons are all man made machines that by their very nature and chosen use will fail from time to time, therefore it is important for one who operates one of these machines to know how to keep it up and running, particularly in a fight. We also train to transition from one weapon system to another, in the event that one is more applicable than the other for the task at hand or the stoppage is too complicated to fix, simple transition drills shifting from long gun or shotgun to the sidearm have been a part of firearms training for countless years. However, in defensive tactics training the idea that an individual may need to transition or shift from one type of technique to another seems to be a fleeting concept. When an officer commits to one technique and fails to achieve minimal response or if the officers tries to transition, with out training, to another techniques and hesitates during the transition he or she has just walked into a tremendously dangerous situation.

As firearms instructors we train students to Transition from weapon to weapon incase of a serious malfunction, but as DT instructors we seldom condition our students to transition from physical technique to technique.

(Courtesy of SG Five)

How do we solve the problem with defensive tactics training? Not so simple an answer, sense many of the programs are firmly entrenched in our community and sense agencies have already committed time and money to them it would be hard to nearly impossible to change now. However, a small tweak of those existing systems is probably the most applicable. I am often asked my opinion as to which DT system or martial arts program I believe is the absolute best for protective occupations and, much like the answer about which gun is best, I usually reply with "which ever one your using". I started martial arts training when I was 12 years old at a very small Shotokan and Kenpo school in North Highlands, California. I was completely in love with this program and thought it would transform me into a bullet proof street fighter, unfortunately and painfully, I was wrong. I studied Shotokan and Kenpo for three years on a near religious basis. Some years later, I moved to Flagstaff, Arizona where I studied Hwarangdo and took up full-contact fighting. I found myself as enamored with this school and what it offered as I was with my first school, and in the controlled world of a regulation ring I found that it alone was not the answer to all my needs. I stayed with Hwarangdo for several years, also dabbling with Tae Kwon Do. A few years later, I enlisted in the U.S. Army and during my basic training and my advanced training I was exposed to "unarmed self-defense" or USD and I told myself that as soon as I graduate, I am heading straight back to the first martial arts school I could find. Fortunately, my first duty station was in the Republic of Korea where martial arts schools are abundant and oddly enough, come in many varieties. During this first tour, and a one year voluntary extension, I was able to study Tang Soo Do, Tae Kwon Do and Kendo on a very intense basis, nearly 3-5 times a week. During this tour, I suffered two broken noses, a number of broken bones in my hands and feet, and a stab wound to my back further leading me to believe that I had not yet found the answer to my problem. Following Korea I returned state side, only venturing outside the borders of the continental United States on occasional TDY assignments. While in the states, I became very dissatisfied with the schools I found. Apparently, after studying in the country of origin for many of my fonder styles spoiled me. I opted to teach martial arts and self-defense part time. I began to notice how I had become as much of the problem as anyone else involved in teaching conceptual self-defense. I realized that my contemporaries and I were offering nothing more than courses comprised of regurgitated meaningless moves. We were taking the very same movements that we had been taught, repackaging them and selling them again. We were doing the age old martial arts marketing ploy of literally walking our students through a series of "self defense" moves on a step by step basis, that by design and application would only work against an opponent who would throw weak, undirected and wild techniques at very slow speeds. I determined that, especially after a number of severe ass beatings, that this type of training was actually causing more harm than good. Unfortunately, this is how a great many martial arts schools and DT gurus teach their trade. An inordinate number of perpetual students spend their hard earned money every year traveling from self-defense class to self defense class, all hoping to learn something new that will make them better, stronger and faster than the bad guys on the street, sometimes they do and sometime they don't. Now that I have strayed from my topic I am going to go back and address what system is the best. There are a great number of styles and systems available, more every year; some are nothing more than a new label on an old box. The real value of a system does not lay in a debate over why grappling in better than kicking or why kicking is better than pressure points. The real issue is how much time is devoted to whatever style you choose in order to gain proficiency and how practical and realistic the training was. If time and energy are committed to the initial training and time and energy are not committed to subsequent training, then all the classes and all the money are more or less wasted. That's it, that's the entire secret over, in the words of my beloved Sunday afternoon Kung Fu flicks, "who's Kung Fu is better!" Today, I teach an extreme bastardized form of about 4 or 5 different styles because I feel that some techniques are better executed and explained by different means.

While most forms of martial art are not taught as a means of conducting actual combat, a great many styles offer us a wide array of techniques that are applicable to our needs and our student's needs.

(Courtesy of Perry Taylor)

Concepts of peak performance

Personally, I feel that discussing the concept of peak performance is highly appropriate for a book of this nature. Peak performance is, or at least should be, something that our students and we should be striving to achieve. It is important to examine and understand the concepts and characteristics of high performance. To understand it means to empower us and our students with the means to control situations and create awareness that is second to none.

The experience of watching a master of combative skills operate is a truly marvelous thing, as the experience both physical and mental aspects that are recognizable and admirable. A visual display of the time spent training and developing is a testament to the individual triumphs and defeats. The meaning and needs of the student who undertakes force application training varies widely with the individual: from financial gain, to glory, to pursuit of a personal goal or to simply the summoning of a higher calling. Regardless of the extrinsic motivation, many times the practitioners of these arts have reported experiencing an intrinsic sense of extreme concentration. Sometimes described as a state of focused energy, a state of enlightened well being and an altered sense of time. These moments when the mind and the body seem to come together as one are difficult to describe, but are considered a beneficial element to the operators overall performance and a significant contribution to the operators mastery of his environment.

Flow or zone is often characterized as a feeling of euphoria and a sense of

everything go right. Endurance and extreme athletes often feel this condition.

(Courtesy of SG Five)

There are many terms used to describe these moments in time when everything seems to be coming together in all the right ways and unfortunately most of them are made in reference to athletic events, but with some careful understanding the viable elements of sports psychology and human performance can be adapted to the arena of force application training. Terms like in the flow or in the zone, are often used to describe the sensation of everything going our way. A period in time where we are so focused, so involved that nothing else matters and nothing can interrupt us.

Long distance runners often refer to this feeling as "runners high" or as a state of euphoria coming on in a very unexpected way and at unexpected times, leaving the runner with a feeling that they have an ability to transcend the barriers of time and space (Sachs, 1984).

The concept of flow or zone comes from two psychological theories, flow theory (Csikszentmihalyi, 175, 1990) and reversal theory (Apter, 1982, 1989). The flow theory specifically states that periods of being in the flow are sporadic in occurrence, but a dynamic period of self-rewarding involvement when it does occur. Flow theory further states that while the zone experienced differently by different people, there are 8 distinct conditions that do occur regardless of the occupation, culture or demographics of those who have experienced it. These conditions are as follows; Clear goals and feedback, balance between challenges and skills, action and awareness coalesce, extreme concentration on the task at hand, a sense of control, a loss of self-consciousness, an altered sense of time and a self-rewarding experience. The reversal theory states that the zone or flow is a metamotivational state or a state where the individual's motives are structured, interpreted and organized within the experience. Additionally, under the reversal theory the individual is believed to experience flow or zone as an optimal relaxing goal.

Wes Doss

Given the serious nature of the various occupations that require skills in force application, combined with the intense stress experienced when under fire and the known potential for suffering that exists, we and our students may not enter a zone or flow state that provides the equivalent feelings of the runners high. The victorious warrior may feel elation and rapturous after the threats are neutralized and the fight is over, but there may be too much intensity during a melee to coexist with the kind of sensations described above. Does this mean that the flow and the zone are fallacies for those of us in the combative crafts? No, not even close. What is does mean is that much like performance athletes in endurance sports who must sustain and endure pain and suffering for long periods of time, we enter a zone or flow state that differs from that which is experienced by others.

For us and our students, particularly those who will potentially operate for extended periods of time, we need to enter into a period of total absorption into the mission or operation and entering into what is referred to as "kairos" time. The Greeks had two words for time; kairos and kronos. Kairos refers to the point when we are so engrossed in our task that we lose our sense of time passing, we are so totally absorbed in the moment that it may actually stretch out for hours (Bolen, 1996). In this form of zone or flow the individual is able to dissociate from pain and extreme conditions. This dissociation, coupled with the absolute focus and the absorption of awareness (Jackson and Csikszentmihalyi, 1999) could be the reason why so many warriors have had the ability to overcome suffering and perform with brilliance, even into death.

Regardless of short term or long term activity or the level of intensity, the ability to enter in the zone or flow is a tremendous asset to all who take their line of work seriously. Unfortunately, these periods of peak performance don't happen often enough or even at the right times. Typically, when it does occur, it usually happens by chance. However, we can do things to increase the likelihood of functioning under peak performance. The primary tool or device to achieve peak performance is already in our possession and has been since before birth. This very specialized appliance is so attuned to developing periods of peak performance that it can actually bring its user into situations that exceed expectation. I am, of course, referring to the ever-popular human mind. The implementation of mental training skills into a force application training program can grossly increase the possibility of falling into a peak performance period on a tremendously more consistent basis. Being in "the zone" or "the flow" means doing more and responding faster than most believe possible.

Mental training skills, like the deep focus used in weight lifting, can increase an

individuals potential of entering a zone or flow state during critical situations.

(Courtesy of SG Five)

For all thoughts in the mind the body has a corresponding reaction, this includes thoughts of things real and things imagined. This is explained easily by examining the feelings and reactions following a bad dream or the startled reaction to a strange noise in the middle of the night. In the realm of force application training and execution great many things are still subject to chance, at best violent situations can be considered predictably unpredictable. The reliability of techniques and equipment are always employed under an air of chance, as is dealing with the attitudes and personalities of others. With so much in our field be a literal crapshoot, why allow our very mindset be another? We all have the power and ability to command, with tremendous authority, our mental faculties. Using it as an effective tool to strip mental barriers that can impede performance.

Law enforcement and military folks spend millions if not more every year buying products that they believe will give them an advantage over their foe. I have already mentioned that the multi-billion dollar industry of "officer survival" preys upon this very human of behaviors, but what I find amazing is that the best edge up on the other guy is in our very head. It has been said countless times that shooting or physical tactics are 90% attitude and 10% ability, and while this is more or less true, why do so many do so little to address it? This saying may be common knowledge, but it is far from common practice. Regardless of the reasoning why, the fact remains that many trainers and training programs are failing to utilize our most precious and powerful resource. This seems very odd since it is well documented that under stress we typically fatigue mentally long before we fatigue physically.

Intense mental and physical conditioning is an absolute must for developing winners of conflict situations. Everyone reacts differently when under fire and without attaching training to the physical and mental parts of the student we leave to much to chance.

(Courtesy of Perry Taylor)

Since the heightened state or zone condition is the center of this discussion is important to note and understand the characteristic of this state. The first element of the zone or flow is relaxation. Considerable research has been devoted to the zone experience in sports and it has shown consistently that optimum performance is best achieved when the individual is in a condition where they are just slightly above their normal, and very individual, state of arousal, not at the extreme end of the spectrum. For several years, athletes would get "psyched up" before competition, but current research is showing that being in a relaxed but energized state of balanced intensity is the most productive. The mind is calm while the body is ready to go. In this mode we feel relaxed but we are able to move with explosive speed and power.

Following relaxation, a sense of confidence is important to the zone experience. This sense of confidence must be strong and must not let minor breakdowns in performance undermine the belief in ones overall ability to perform. Confidence carried on the inside but exuded outwardly. This sense of confidence is paramount to winning and most importantly it must be based on reality and should never be false. Based on training and conditioning, you or your students should expect to win, never hoping or wishing to succeed, but an honest sense that they will win.

Closely related to confidence and relaxation in the pursuit of the zone is the ability to become completely absorbed into the situation. The ability to focus or concentrate on the task at hand and ignore distraction is critical for success and is a preeminent condition of the zone. This focus is similar to the focus possessed by a child when completely absorbed with a new toy, not allowing anything or anyone to interfere.

In the zone, things seem to occur automatically and effortlessly. This sensation makes all moves and actions seem smooth as silk and as simple as tying your shoes. The mind and the body are flowing in such tremendous unison that it almost makes one feel like you are watching someone else perform the task. The effortless grace and ease make even complex tasks see simple and uncomplicated by unnecessary thoughts and emotions. Deliberate actions occur without protest and without consent, almost as if we were on automatic pilot. However, instead of being guided by an external mechanical guide we are in complete control. As thoughts and decisions are constructed, they are executed immediately and without regard to our emotions. This sense of command over the situation is strong and allows for great things to be accomplished.

As you read this, I'm sure that a few of you are saying something to the effect of "That's great, but how do we condition the zone?" Well, I'm glad you asked, after all, the only bad question is the one that's not asked. In a violent encounter, a tremendous amount of stress overwhelms the participants (a severe understatement). A phenomenal portion of that stress can be attributed to the participants desperately trying to control the uncontrollable factors of the encounter. When this emphasis is placed on those uncontrollable details, the combatant has a tendency to fail or choke in the execution of desired skills. Initially we [instructors] need to emphasize conditioning our students to focus on the processes required to win an encounter, while ignoring the outcome of winning or losing. Simply put this is the idea of coaching them to focus on what they have to do to win and not dwelling on winning.

The instructor plays an integral part in getting students to focus on the various

processes to reach success, rather than dwelling on the outcome.

(Courtesy of SG Five)

If our students are provided with accurate information in an understandable form that helps them better understand the relationship between their level of stress and their ability to perform, we will be empowering our students to better cope with the rigors of desperate situations. If a soldier or police officer can "read" their own stress about their performance and can tell the difference between good stress, bad stress, and no stress, they can gain a mental advantage placing them in a position to understand. If the level of understanding is elevated then the individual can mitigate, with great success, their arousal level before it climbs sky high and takes over. Additionally, conditioning our students to look at adversity in a different light will also empower them with the ability to work through many problems. Simply, this means having our students look far past the obvious of a situation and find the advantage in the disadvantage, help them see that crappy weather, strange locations and fatigue can often be exploited to their advantage.

Practice makes perfect? Hardly, perfect practice makes perfect. Without doubt, it is the quality of the practice that is ultimately responsible for how much our students gain from practice and ultimately how well they perform under pressure. When we are able to incorporate realistic elements into our training, we are helping to condition our students to adjust and cope with the actual pressure of a real situation. The more our training resembles an actual high risk situation, the less chance our students will have of coming apart under pressure, just like how professional football players scrimmage during practice on a regular basis. If we have students who have problems with the application of certain techniques, we can place greater emphasis on these problems during our simulation training.

Solid simulation training that replicates thestress and dangers of a real world

encounter is priceless.

(Courtesy of SG Five)

When a soldier, police officer or protective professional is threatened with consequences when they don't perform well they run an incredible risk of having that threat come up during pressure, causing them to break down. Threats only serve to distract the operator from the task at hand and cause undue worry about the consequences. Focusing on the 'what ifs' and the 'who got sued for what's' is the last thing we want our folks to be doing. Instead, challenge your students. Provide your students a message, one that tells them that they can do it and that you believe in them. Our students will often rise to the challenge, but will often respond poorly and inconsistently when faced with a threat. Finally, we need to focus our students to perform with optimum performance under pressure. I know that this is essentially what this manual is all about, but it is very relevant to bring it up again here. Most stress related to performance problems is a direct result of faulty concentration. The operator that gets psyched out or intimidated does so because he or she is having problems focusing or is focusing on the wrong things. We need to help our students concentrate on specifically what they have to do to perform well. We need to teach them to control themselves especially their sight and their hearing. The operator needs to only look and listen to those things that keep them composed and performing their very best.

Once again, the self-proclaimed Expert

The list of tragedies occurring in law enforcement and military environments is literally endless and while the finger of blame is often pointed at someone in particular these tragedies stand as a testament to the procedural, tactical and training failures of the agency. More often than not, these tragedies could have been avoided had the ill-fated participants been properly trained and assessed prior to deployment. A great number of administrators and instructors find it difficult to look at their programs critically, identify flaws and accept that they may be providing an insufficient product. One solution is to go to sources outside of the agency, but because of limited budgets outside consultation and training is often not possible. It is interesting to note with an ironic twist that the cost of appropriate and relevant training is far less expensive than the cost of a lawsuit and far less traumatic than a funeral.

Despite the plethora of journals, books and research devoted to education and training, two fundamental questions are still asked and yet often ignored: (1) how well are students learning? And (2) how effective are we teaching? The critical assessment of the training process and an understanding of our impact on our students is paramount.

Through close observation of our classes and the collection of frequent feedback we can learn a tremendous amount about how adults learn and how they respond to different teaching techniques. The assessment of the training process holds some significant information for both the individual instructor and the training facility, as information gathered can be used to regain focus on the curriculum and help make the training process more productive. However, a great number of instructors have gone on for years and exposed to countless numbers of students assuming that students were learning solely by the virtue of them teaching. Far too often students have not learned as much as was expected. There exists gaps, often huge gaps, between what was taught and what was learned (Angelo, Cross, 2000). By the time these gaps are acknowledged, it's usually too late to do anything about it. Now I am sure that there are those who will read this and say to themselves that they solicit for feedback from students regularly in the form of class evaluation forms and while this is a form of assessment, it is not a measure of how well the student assimilated the material. In most classes, commercial or otherwise, there exist two types of students; those who came for the training and those who came for a piece of paper that says they had the training. Typically, both will tell you that the class was great so long as they either get to go or get the certificate. Class evaluations, when they're honest, are only reflections of how the student felt about the course and provide no insight as to how well you conveyed the information or how well the student absorbed it. The instructor, in the form detailed proficiency exercises, conducts the class evaluation that we need to be concerned with. Some method that challenges the student to use the material taught that is fair and objective. However, while the standard practices is to test students at the end of the training session, this is far too late to allow for any corrections to be made. This especially true with the short term classes as well as long term college and academy classes. Consider the assessment process as a long-term continuous process that occurs parallel to the training itself. Never forget that the quality of learning is directly related to the quality of the teaching and one way to improve the quality of the learning is to enhance the teaching. At the very start of a course make your goals and objectives crystal clear and then gather continuous comprehensible feedback as to the extent that the goals and objectives are being met. Consequently, in order for students to improve even those attending short-term classes, they too need appropriate and focused feedback as often as possible. Remember that feedback and reinforcement are critical components to the learning process, without either it is difficult to support a desired response or know that something has been done wrong.

Documentation

As I bring this book to its conclusion, I think it is only appropriate to consider the idea of winning full circle. At the beginning and conclusion of any training program, we find ourselves faced with documentation; lesson plans, pre-tests, quizzes, post tests, evaluations and after action reviews. Is all this necessary? Absolutely, yes! There exists no good reason why training should not be documented in black and white, both our intent and our results.

One of our primary roles as instructors is to design and implement blocks of instruction. Instructors in all disciplines prepare plans that aid in organization and in the delivery of the material. These plans tend to vary widely in style and in the level of detail. Some plans are elaborately detailed while others are very brief. Regardless of the format, all instructors must make wise decisions about their material, classroom strategies and the methods they will use to teach. Instructors, by both necessity and from liability concerns, need much more than a vaguely written list of topics. Effective instructors must understand that the lesson plans that they develop help provide direction toward the attainment of very specific objectives, as well help shelter their respective organizations from the pains of civil litigation. While even the most specific lesson plans leave some room for free expression on the part of the instructor, a core of material that must be covered needs to be the basis of the plan.

Training records and after action reports, including student evaluations, also need to be maintained. These records represent not just documentation

223

of material covered, but also help critically evaluate the success of the class and what needs to changed. While obviously not "experts" our students do provide us with some valuable information, so long as we take it. Many times, and I will mention no names or academies, I have witnessed instructors throw away evaluations that were less than stellar, refusing to acknowledge a potential problem or even consider the need for improvement. I find this to be a shameful travesty of our occupation and a grave disservice to our students. Folks like this need to step back and genuinely consider why they became instructors in the first place and maybe consider a different line of work. I will be among the many who will openly admit, "I make mistakes". I am human and in being human, I am prone to falling on my posterior once and a while. Sometimes I realize it and sometimes I don't, sometimes it takes an honest evaluation for me to realize it. Hey guys and gals, accept it, deal with it, fix the problem and drive on! We are not here as instructors to impose our idiosyncrasies on others. We are here to help good people win and go home to their families at the end of the day.

A final word on documentation, I remember a story, one I am sure many of you have heard and may have actually witnessed. An academy class on some type of arrest procedure is being conducted. The instructor decides to take a smoke break and puts a video on for the class. Sometime later after the class graduates, one of the class members decapitates a suspect with a shotgun while applying handcuffs. During the subsequent investigation and court hearings, the officer declares that he was taught to handcuff with one hand while managing a shotgun with the other when he attended the academy. The academy training coordinator and the instructor are brought into court and asked about the subject material, both deny having any knowledge of the alleged training. At some point in time, the video is introduced into the trial and viewed by the members of the court. The video was about some other form of arrest tactic, but in a series of scenes, blurred and in the far background it could be seen a figure holding a shotgun at the base of a man's neck while he applied handcuffs. This was the point where the fat lady started to sing! The moral of the story is be careful of the material you provide, document everything accurately and prepare your documentation with the potential of having to defend it in court.

Conclusion

In meeting the challenge of preparing the protectors of our society to win violent confrontations in the world of the 21st century, the various instructors and institutions are faced with the task of rethinking programs and curriculum to take advantage of the literal profusion of training resources as well as specific research into how adult humans learn. Our goal as instructors must be without failure to create a new generation of protectors, a new and more effective generation that is more creative and resourceful in their thinking and their application of their trade.

Trainers and especially training institutions should be considered bastions of higher training and education. Those involved in this trade must acknowledge the needs of their students and develop curriculum that meets those needs. The average police officer or soldier does not enter these institutions as an empty palate or clean slate. These professionals bring with them a lifetime of experience. They are at an operational stage in their lives and in their way of thinking. Education for the modern adult student is, or at least should be, a highly individualized experience built on a life-long pursuit of being the best and being a winner.

Only with credible and relevant practice can timely and appropriate performance is

attained.

(Illustration courtesy of Mark Burnell)

Progressive instructors who understand and recognize the needs of their students should reject the assumption that training and education are nothing more than a bundle of senseless, time filled, brass building courses whose usefulness desists at the end of the training session. In our world of force application training there exists a severe training crisis and as a result, how society and our peers view training. The stereotypical classroom is a place where the students sit quiet and listen to the teacher. This concept represents nothing more than a dysfunctional situation, especially when it comes to skills that are designed to save or take lives. The stiff, proper, black and white, sterile classroom of the past may have met the needs of the students of earlier times, but it does little to address the training needs of the modern student of force application. While most agree that force application training is best dealt with as a hands-on type function, the concept of teaching is always going to be dictated by the various administrative perceptions of how best training should occur. However, we [instructors] are the only folks in the position to initiate the change.

Far too often, information is treated like a source of power and prestige, but information is not the sole domain of the instructor. It is available to anyone who knows how to find it and use it. The amount of training resources and the Internet has increased the availability of information to the individual learner considerably over the last 10 – 15 years, far beyond anyone's imagination. This upsurge of available wisdom is far more than any mere mortal could manage and dictates that committing all the information to memory is an utterly impossible task. It is now imperative for our students, regardless of their specific occupation, to become users of information in lieu of merely being collectors. We, as instructors, must stop viewing ourselves as the infamous "fountains of knowledge" and begin to see ourselves as " developers of knowledge" and work more to help the student learn and ultimately help the student win.

Today and in the future, warrior-training programs need to delve into the ways that humans learn and how they can tailor the information to meet the specific needs of the student. In meeting the challenge of developing warriors in a world of growing complexity, instructors are faced with the task of integrating reality and functional stress into their courses. As warriors ourselves, we must be able to prepare our students to take our place in a dynamic world of constant change. The concepts reflected in this text have the potential of helping others make dramatic changes in the lives of others, but it is important to remember that this is only a small scratch on the surface of a much larger issue. For any program or individual instructor to be effective in setting students on the path to winning they must go beyond the narrow focus of the standard or the norm venturing far outside of the box.

"The art of teaching is the art of assisting discovery"

-Mark Van Doren

"Train to win."

-Wes Doss

References

Adams, J.A. (1987). Historical Review and Appraisal of Research on the Learning, Retention, and Transfer of Human Motor Skills. Psychological Bulletin, 101, 41-74

Allport, G.W. & Postman, L.J. (1947), The Psychology of Rumor, New York: Holt

Ames, C. & Ames, R. (1989). Research in Motivation in Education, Vol 3. San Diego: Academic Press

Anderson, L. & Krathwohl, D. (2001). A Taxonomy for Learning, Teaching and Assessing: A Revision of Bloom's Taxonomy of Educational Objectives. New York: Longman.

Bacon, S.J. (1974), Arousal and the Range of Cue Utilization. Journal of Experimental Psychology, 102, 81-87

Bartlett, F.C. (193), Remembering: A Study in Experimental and Social Psychology, London: Cambridge Press

Wes Doss

Bartlett, F.C. (1958). Thinking. New York: Basic Books

Berlyne, D. (1960). Conflict, Arousal, and Curiosity. New York: McGraw Hill

Broadbent, D. (1958). Perception and Communcation. London:Pergamon Press

Brookfield, S. (2001). Adult Learning: An Overview. http://www.nl.edu

Chang, T.M. (1986), Semantic Memory: Facts and Models. Psychological Bulletin, 99 (2), 199-200.

Collins, A.S. (1978). Common Sense Training. San Rafael, CA: Presidio Press

Desmedt, J. (2000). Fixing the Use of Force Machine. http://pss.cc/nuc-art.htm

Deterline, W.A. (1962). An Introduction to Programmed Instruction. New York: Prentice-Hall

Dixon, N.F. (1971). Subliminal Perception: The Nature of a Controversy. New York: McGraw-Hill

Ellis, W.D. (1938). A Source Book of Gestalt Psychology. New York: Harcourt, Brace & World

Gagne, R. (1962). Military Training and Principles of Learning. American Psychologist, 17, 263-276

Gagne, R. (1985). The Conditions of Learning (4th ed.). New York: Holt, Rinehart & Winston.

Gibson, E. (1969). Principles of Perceptual Learning and Development. New York: Appleton

Glynn, S.M. & DiVesta, F.J. (1977). Outline and Hierarchical Organization for Study and Retrieval.
Journal of Educational Psychology, 77(2), 137-148

Good, K.J. (2000) Let There Be Light, http://www.surefire.com

Greenwald, A.W. (1992). New Look 3: Unconscious Cognition Reclaimed. American Psychologist, 47, 766-779

Guthrie, E.R. (1935). The Psychology of Learning. New York: Harper

Guthrie, E.R. (1938). The Psychology of Human Conflict. New York: Harper

Haber, R.N. (1969), Eidetic Images, Scientific American, 220, 36-44

Humara, M. (2001). The Relationship Between Anxiety and Performance: A Cognitive-Behavioral Perspective. http://www.athleticinsight.com

Khalsa, D.S. (1997), Brain Longevity. New York: Time-Warner Books

Knowles, M. (1975) Self-Directed Learning. Chicago: Follet

Knowles, M. (1984). Andragogy in Action. San Francisco: Jossey-Bass

Kolb, D.A. (1984). Experiential Learning. Englewood Cliffs, NJ: Prentice-Hill

Lave, J.(1988) Cognition in Practice: Mind Mathematics and Culture in Everyday Life. Cambridge, UK: Cambridge University Press.

Luce, G.G. (1971), Body Time. New York: Random House

Norman, D. (1967). Memory and Attention. New York: Wiley

Mandler, G. (1984). Mind and Body. New York: Norton

McClelland, D. (1985). Human Motivation. Glenview, IL:Scott, Foresman

Reiss, S. (2001) Why America Loves Reality TV. Psychology Today

Rumelhart, D.E. (1980), Schemata: The Building Blocks of Cognition. In R.J. Spiro, B. Bruce, & W.F. Brewer (eds.), Theoretical Issues in Reading and Comprehension. Hillsdale, NJ: Erlbaum

Schinke, R.J. & Tabakman J. (2001) Reflective Coaching Interventions for Athletic Excellence,

http://www.athleticinsight.com

Siddle, B. (1995), Sharpening the Warriors Edge, PPCT Research Publications

Squire, L.R. (1986), Mechanisms of Memory. Science, 23, 1612-1619

Sticht, T.G.(1988). Adult Literacy Education. Review of Research in Education, Vol 15.

Washington DC: American Education Research Association

Suarez, G. (1997), The Tactical Advantage, Paladin Press

Treisman, A.M. (1960), Contextual Cues in Selective Listening. Quarterly Journal of Experimental Psychology, 77, 206-219

Trungpa Chogyam, (1988), Shambhala: The Sacred Path of the Warrior, London: Shambhala Publications

Williams, F.D. (1990). SLAM: The Influence of S.L.A. Marshall on the United States Army. U.S. Army, TRADOC, Fort Monroe, VA.

About the Author:

Wes Doss is currently one of the owners of Khyber Interactive Associates, an emergency response training and consulting firm and the Executive Vice President of SG Five, an international law enforcement and military training organization. Additionally, Wes has served as the Senior Firearms Instructor for the Sig Arms Academy in New Hampshire. Formerly, a Patrol Sergeant with the Mohave County Sheriffs Office in Northwestern Arizona, he served 10-years as a detective, training officer, and patrol deputy. Wes supervised his agencies firearms training unit and was the team leader and training coordinator for the 17 man tactical operations unit (TOU). Prior to embarking on his civilian law enforcement career, Wes served for 7 years in the U.S. Army, and is currently active with the U.S. Army Reserve.

Wes holds both a B.S. and an M.S. in Criminal Justice Administration and is pursuing a doctorate in psychology. Wes holds numerous certifications from the U.S. Army, U.S. Marine Corps, Arizona POST, The Smith & Wesson Academy, The NRA, and The Federal Emergency Management Agency (FEMA). Wes is also deeply involved the martial arts, having studied a variety of different styles for over 20 years, holding black belt/instructor ratings in Hapkido and Hwa Rang Do.

Wes is a faculty member for the local community college and holds teaching credentials through the State of Arizona. Wes is active is instructing both general curriculum courses and criminal justice courses. Recently Wes has been getting involved in the writing and general development of curriculum for the college and the local law enforcement training academy.

Wes has also become involved in the writing field, having articles published in ASLET's magazine *The Trainer*, SWAT magazine, and the NTOA's publication *The Tactical Edge*. Wes is an active member in the National Tactical Officers Association (NTOA), The American Society of Law Enforcement Training (ASLET), The International Association of Law Enforcement Firearms Instructors (IALEFI), The Military Police Regimental Association (MPRA), The United States Deputy Sheriffs Association (USDSA), The National Rifle Association (NRA), and The Arizona Homicide Investigators Association (AHIA).

CPSIA information can be obtained
at www.ICGtesting.com
Printed in the USA
FSOW02n1613290416
19884FS